Additive Manufacturing in Optics and Photonics

Fabrication and applications

Online at: https://doi.org/10.1088/978-0-7503-6428-7

IOP Series in Emerging Technologies in Optics and Photonics

Series Editor

R Barry Johnson, a Senior Research Professor at Alabama A&M University, has been involved for over 50 years in lens design, optical systems design, electro-optical systems engineering, photonics, and deep learning. He has been a faculty member at three academic institutions engaged in optics education and research, has been employed by a number of companies, and has provided consulting services.

Dr Johnson is an IOP Fellow, an SPIE Fellow and Honorary Member, an Optica Fellow, and was the 1987 President of SPIE. He has served on the editorial board of *Infrared Physics & Technology and Advances in Optical Technologies*. Dr Johnson has been awarded many patents, has published numerous papers and several books and book chapters, and was awarded the 2012 Optica (OSA)/SPIE Joseph W Goodman Book Writing Award for *Lens Design Fundamentals, Second Edition*. Until 2024, he was a perennial co-chair of the annual SPIE Current Developments in Lens Design and Optical Engineering Conference.

Foreword

Until the 1960s the field of optics was primarily concentrated in the classical areas of photography, cameras, binoculars, telescopes, spectrometers, colorimeters, radiometers, etc. In the late 1960s optics began to blossom with the advent of new types of infrared detectors, liquid crystal displays (LCDs), light emitting diodes (LEDs), charge coupled devices (CCDs), lasers, holography, and fiber optics along with new optical materials, advances in optical and mechanical fabrication, new optical design programs, and many more technologies. With the development of the LED, LCD, CCD, and other electro-optical devices, the term 'photonics' came into vogue in the 1980s to describe the science of using light in the development of new technologies and the operation of a myriad of applications. Today optics and photonics are truly pervasive throughout society and new technologies are continuing to emerge. The objective of this series is to provide students, researchers, and those who enjoy self-education with a wide-ranging collection of books, each of which focuses on a topic relevant to the technologies and applications of optics and photonics. These books will provide knowledge to prepare the reader to be better able to participate in these exciting areas now and in the future. The title of this series is *Emerging Technologies in Optics and Photonics*, in which 'emerging' is taken to mean 'coming into existence', 'coming into maturity', and 'coming into prominence'. IOP Publishing and I hope that you will find this series of significant value to you and your career.

A full list of the titles published in this series can be found at: https://iopscience.iop.org/bookListInfo/emerging-technologies-in-optics-and-photonics.

Additive Manufacturing in Optics and Photonics

Fabrication and applications

Ricardo Oliveira

Instituto de Telecomunicações and University of Aveiro, Campus Universitário de Santiago, 3810-193 Aveiro, Portugal

Nuno Valente

Instituto de Telecomunicações and University of Aveiro, Campus Universitário de Santiago, 3810-193 Aveiro, Portugal

IOP Publishing, Bristol, UK

ISBN   978-0-7503-6428-7 (ebook)
ISBN   978-0-7503-6431-7 (print)
ISBN   978-0-7503-6430-0 (myPrint)
ISBN   978-0-7503-6429-4 (mobi)

DOI   10.1088/978-0-7503-6428-7

Version: 20250901

IOP ebooks

British Library Cataloguing-in-Publication Data: A catalogue record for this book is available from the British Library.

Published by IOP Publishing, wholly owned by The Institute of Physics, London

IOP Publishing, No.2 The Distillery, Glassfields, Avon Street, Bristol, BS2 0GR, UK

US Office: IOP Publishing, Inc., 190 North Independence Mall West, Suite 601, Philadelphia, PA 19106, USA

Contents

Foreword

In the history of science and technology, few tools have offered as much promise to transform fabrication as additive manufacturing (AM). Originally heralded for its ability to streamline prototyping, AM has matured into a powerful platform for production, which can enable a new paradigm of custom, complex, and high-precision components across diverse industries. Among these, optics and photonics stand out as fields that demand unmatched fidelity, surface quality, and material control. For decades, these requirements have relegated optical component fabrication to a domain of artisanal craftsmanship and specialized infrastructure. Today, that exclusivity is rapidly dissolving. At the convergence of advanced photonics and next-generation fabrication, *Additive Manufacturing in Optics and Photonics: Fabrication and Applications* provides a compelling, comprehensive roadmap for what is now possible, and what is just beginning.

Dr. Ricardo Jorge Figueiredo Oliveira and Mr. Nuno Fonseca Valente, researchers at the Instituto de Telecomunicações and University of Aveiro, have crafted a work of remarkable depth and clarity. This book is not merely a survey of techniques or a catalog of demonstrations. Fundamentally, it is a forward-looking blend: a treatise on how the evolving capabilities of AM are poised to revolutionize the design, prototyping, and deployment of optical and photonic systems. Their message is clear; additive manufacturing is not an auxiliary method for optics, but it is an enabling one.

The text begins with a detailed and technically rich exploration of the full landscape of AM technologies. Techniques such as stereolithography (SLA), selective laser sintering (SLS), fused deposition modeling (FDM), material jetting (MJ), and more advanced methods like two-photon polymerization (TPP) and continuous liquid interface production (CLIP) are not just described but critically evaluated in the context of optics. The authors articulate the nuances of each method with precision by highlighting aspects such as resolution limits, layer thickness, surface finish, post-processing needs, material compatibility, and optical transparency. Of particular importance is the book's attention to the critical junction between AM hardware capabilities and the stringent optical performance metrics that photonics applications demand.

Yet the true strength of this volume lies in its applied orientation. The second part of the book transitions from processes to products, and from fabrication science to functional optics. Here, the authors offer an impressive survey of 3D-printed optical components, ranging from classical lenses to advanced freeform optics, reflective surfaces, optomechanical mounts, optical fibers, amplitude masks, and integration-ready optical setups. This is where the promise of AM crystallizes. No longer constrained by conventional machining geometries or the cost barriers of injection molds, researchers can now explore nontraditional optical surfaces, tailor components to emerging device architectures, and iterate on designs in-house at a pace never before possible!

The chapter on lenses is particularly illuminating. The authors describe how AM enables not only the production of basic spherical or aspherical elements, but also sophisticated freeform lenses, i.e., those that integrate multiple optical functions in a single element. These were once relegated to high-end defense or aerospace applications due to manufacturing limitations; today, with MJ or TPP printing, they can be fabricated with remarkable fidelity and minimal post-processing. The discussion includes comparisons of wavefront error, roughness, and transmission across techniques, presenting the reader with a realistic picture of what is achievable today and where key challenges remain.

Surface quality, long considered the weakness of 3D printing for optics, is addressed with refreshing technical honesty. The authors don't gloss over the limitations. Instead, they provide detailed discussions of UV post-curing, solvent polishing, thermal annealing, and other finishing strategies, offering concrete paths for transitioning printed prototypes into functional optical-grade components. In this regard, the book is both aspiring and practical, and a guide for pushing the frontier, but always grounded in rigorous experimentation and methodical evaluation.

Equally noteworthy is the book's treatment of photonic applications beyond static optics. The authors introduce 3D-printed fiber holders, phase masks, and optomechanical components integrated directly into experimental systems, thereby demonstrating how AM isn't simply a tool for component production, but a strategy for system-level innovation. These examples underscore one of the book's most important themes: AM is not just making existing optics cheaper or faster; it is enabling designs that were previously inconceivable.

Another distinguishing feature of this work is its accessibility across disciplines. Optical engineers, materials scientists, additive manufacturing specialists, and applied physicists will all find the content compelling. Graduate students will appreciate the clarity of explanation and breadth of references, while industry practitioners will benefit from the detailed breakdowns of fabrication constraints, material considerations, and use-case studies. The structure of the manuscript reflects a thoughtful pedagogical strategy that moves from theory and process into application, with careful transitions and well-organized data that support learning and reference.

This book also arrives at a pivotal moment. As the photonics industry grapples with the rising demand for compact, high-performance, and customized devices in fields such as quantum technologies, biomedical imaging, autonomous sensing, and space systems, the need for agile and scalable manufacturing has never been more urgent. Additive manufacturing presents a compelling solution, one that aligns with the increasing miniaturization, integration, and specialization of optical systems. With the accessibility of AM tools to everyone through lower-cost desktop SLA printers, open-source design libraries, and advances in printable optical resins, this trend is further accelerated.

As someone who has witnessed the evolution of both optical engineering and advanced manufacturing for over 50 years, I view this book as more than a technical reference. It is a signal to the community to rethink boundaries. What we once considered as fabrication constraints may soon be design parameters. What used to require a cleanroom, CNC machine, or custom tooling can now begin as a digital model, printed overnight, and tested the next morning. With the guidance provided in this volume, that vision becomes actionable.

Dr. Oliveira and Mr. Valente have given the optics and photonics community a gift: a meticulously researched, clearly articulated, and profoundly insightful road-map for engaging through additive manufacturing. It reflects not just the state-of-the-art, but the momentum and direction of the field. For those seeking to build the next generation of more compact, more functional, more accessible optical systems, this book is an excellent companion.

Whether you are a researcher probing the edge of photonic integration, an engineer developing novel sensors, or an educator preparing students for the future of optical fabrication, *Additive Manufacturing in Optics and Photonics* will serve as an essential guide. In these pages lies a glimpse of the optical systems of tomorrow, and the tools to begin building them today. As additive manufacturing continues to evolve, so too will its impact on optics and photonics. The insights captured in this volume will help ensure that readers are not only prepared for this future but also equipped to lead it.

R. Barry Johnson, D.Sc., FInstP, FOSA, FSPIE, HonSPIE
Senior Research Professor
Institute of Physics Publishing Series Editor, *Emerging Technologies in Optics and Photonics*
Past-President SPIE – The International Society for Optics and Photonics
Department of Physics, Chemistry and Mathematics
and
Department of Electrical Engineering and Computer Science
College of Engineering, Technology, and Physical Sciences
Alabama A&M University
Huntsville, Alabama
USA
August 2025

Preface

Additive manufacturing (AM), also known as 3D printing, is rapidly transforming both industry and academic research. In optics and photonics, AM offers exciting possibilities: from fabricating custom freeform components to reducing material use, cutting production costs, and enabling on-demand prototyping. These features have made AM a powerful tool for designing and building novel optical elements with unique shapes, materials, and functions.

This book is intended to guide graduate students and early-career researchers through the current landscape of AM for optics and photonics. It explores both the theoretical foundations and the practical considerations necessary to understand how these technologies are utilised in the fabrication of optical devices.

Chapter 1 introduces the main AM techniques relevant to photonics, such as stereolithography (SLA), selective laser sintering (SLS), digital light processing (DLP), binder jetting, two-photon polymerisation (TPP), and direct laser writing (DLW). The chapter compares their working principles, resolution limits, compatible materials, and typical applications. Chapter 2 focuses on 3D-printed free-space optical components. With AM, lenses, beam shapers, integrating spheres, and even parts of complex optical instruments can now be fabricated directly, offering unmatched flexibility for experimental design and prototyping. Chapter 3 explores how AM is used to fabricate optical waveguides. By replacing traditional methods like fibre drawing with customizable 3D-printed designs, researchers can now create waveguides with specialised geometries for splitters, filters, and other integrated components, without relying on expensive manufacturing equipment. Chapter 4 introduces advanced techniques such as TPP, which enables the production of complex structures with sub-micrometre resolution and minimal surface roughness. Applications include the development of micro-resonators, photonic crystals, diffractive elements, and quantum photonic structures. The chapter also provides an overview of current commercial systems and materials that support nanoscale photonic design.

We hope this book provides a strong foundation and sparks curiosity in how AM is enabling a new era of optical design and innovation. Additionally, our aim is to support both newcomers and experienced researchers by providing a clear overview of the potential and limitations of 3D printing technologies applied to optics and photonics.

Acknowledgments

The authors would like to sincerely thank Dr R Barry Johnson (Senior Research Professor at Alabama A&M University, Normal, AL, USA) for his valuable guidance, mentorship, and continued support. His encouragement and vision have greatly contributed to the development of this work, and it has been a privilege to contribute to his book series. The authors acknowledge Fundação para a Ciência e a Tecnologia (FCT) for its financial support through IT Base Funding with project reference UID/50008/2023 IT. The authors also acknowledge FCT funding for the AM-OPTICAL projects, FCT references: PTDC/EMEEME/4593/2021 and CEECIND 2021.01066.

About the authors

Ricardo Oliveira

Ricardo Oliveira received his PhD in physics engineering from the University of Aveiro, Portugal, in September 2017 and his MSc in biomedical engineering from the University of Coimbra, Portugal, in 2010. Throughout his career, Ricardo has contributed to 21 research projects focused on fibre optic technologies, particularly in the areas of sensing, devices, and communications.

Ricardo is currently a researcher at the Instituto de Telecomunicações in Aveiro, where he serves as principal investigator (PI) and co-PI of different projects focused on AM, multi-core fibre applications, polymer optical fibres, and sensors.

Since 2018, Ricardo has held the position of Invited Professor in Applied Optics at the University of Aveiro's Physics Department. Ricardo has authored over 40 papers in leading international journals and more than 40 papers in international conference proceedings, including five invited contributions. He is also the first author of books and book chapters. His research interests include short- and long-period fibre gratings, fibre optic sensors, optical communications, polymer optical fibres, microstructured fibres, additive manufacturing, fibre post-processing, 3D printing, and two-photon direct laser writing.

E-mail: oliveiraricardo@ua.pt

Nuno Valente

Nuno Valente received his MSc degree in physics engineering from the University of Aveiro, Portugal, in July 2023. He then worked as a researcher on a nationally funded project (FOPE-COMSENS—PTDC/EEI-TEL/1511/2020) at the Instituto de Telecomunicações in Aveiro, Portugal. His areas of expertise include fibre optic sensors and components, such as fibre interferometers and fibre grating technology. His expertise extends to the field of freeform AM of optical waveguides and components. Currently, Nuno works as a calculation and simulation engineer at Bontaz group, where he performs structural analysis using ANSYS software and conducts simulations in fluid dynamics and electromagnetics.

E-mail: valentefnuno@ua.pt

IOP Publishing

Additive Manufacturing in Optics and Photonics
Fabrication and applications
Ricardo Oliveira and Nuno Valente

Chapter 1

Additive manufacturing: an overview

1.1 Introduction—3D printing: process chain overview

In recent years, additive manufacturing (AM) has gained increasing attention due to its ease of fabrication and high level of detail. AM has been explored in several applications, including the aerospace industry, medical implants, robotics, and, more recently, optics. In optics, the opportunities presented by AM have only become feasible through the use of transparent materials and the ability to print at low resolutions, ranging from the microscale for widely available, low-cost 3D printers to the nanoscale for the most advanced printers currently available on the market. This capability has enabled the production of various optical components, allowing the scientific community to explore innovative and exotic methods for guiding and manipulating light at the micro- and nanoscale.

The adoption of AM in optics presents numerous opportunities for developing optical components with intricate free-form designs that were once considered impossible to produce due to technological limitations or high production costs. Furthermore, AM enables the rapid prototyping of optical components, accelerating the production and testing of new optical designs. This efficiency paves the way for faster development of market-ready products than ever before. As a result, researchers can develop multiple products and quickly transition from digital prototypes to real-world applications. This ease of production is invaluable for both research applications and the industry. Sustainability is also another advantage to consider with AM, as material is applied only where necessary, thereby minimising waste. This contrasts with current technologies such as the conventional milling technique, which generates a significant amount of waste during production.

The continuous use of AM in optics demonstrates that this technology is set to transform the field. Therefore, a comprehensive understanding of the impact of AM on optics requires an analysis of the various existing AM techniques.

AM processes differ from conventional subtractive fabrication methods, which often involve labour-intensive grinding, polishing, and moulding processes. At its

doi:10.1088/978-0-7503-6428-7ch1   1-1

core, AM is centred around the transformative concept of building three-dimensional components by adding layers one on top of the other. In AM, the printing process begins with the creation of a digital model, typically generated using computer-aided design (CAD) software or by making a 3D scan of a physical object. This serves as the virtual blueprint for the physical component. Then, the 3D model is converted to a standard tessellation language (STL) format, which represents the surface information of the object as a mesh of triangles, with sizes that depend on the resolution of the object to be printed; the smaller the triangle size, the smoother the resulting 3D part. Later, the STL file is imported into slicer software, which allows for the inclusion (if necessary) of pillar supports on object surfaces that are prone to falling due to gravitational force. More importantly, the slicer software enables the user to slice the model into layers that are parallel to the printing platform. Information about the layer thickness, printing velocity, and other settings is stored in a file containing a set of G-code instructions, which is then transferred to the 3D printer. This printer gradually adds material until the intended component takes its final shape.

After the printing process is complete, the 3D-printed components must be removed from the printing bed or build platform. Then, depending on the 3D printing method, they need to undergo various post-processing techniques, which may involve washing, support removal, post-curing, thermal treatment, and chemical or thermal treatments for surface roughness. In the case of vat polymerisation techniques (in other words, 3D printing processes that involve printing objects directly from liquid photopolymerisable resins), the fresh 'green' 3D-printed component needs to pass through a series of procedures [1]. The most important aspect relates to cleaning the uncured liquid resin that coats the freshly 3D-printed parts. For this, the components are rinsed with a cleaning agent. Isopropyl alcohol (IPA) is frequently used, and the process may involve immersing the object in alcohol for a specific duration. Ultrasonic cleaning can effectively remove liquid resin stuck in tight spots, revealing intricate features of the object. It is important to note that polymers have liquid absorption capabilities; therefore, the washing time should be brief to preserve the raw properties of the polymer material. Typically, the washing process lasts for 2–3 min. For conservative reasons, some studies report the use of short heat treatments at low temperatures to evaporate the IPA that may be absorbed and trapped in the polymer matrix. The next step may involve removing supports with the aid of a wire cutter. At this stage, the 3D object is not completely polymerised; thus, an ultraviolet (UV) post-processing step is used to enhance the resistance and stiffness of the printed components [2]. Typically, this can be achieved in a 3–6 min process at a 3D printing curing station. Depending on the thickness of the 3D-printed object and the high attenuation of polymers in the UV wavelength range, the post-curing process may take longer. UV radiation cannot fully penetrate thicker objects, leading to anisotropic polymerisation distribution, with side effects manifesting as shape deformation and/or improper mechanical, optical, thermal, and other properties.

Depending on the material to be printed and the 3D printing technique used, the sintering post-processing technique may also be applied. Examples of materials that fall into this category include ceramics, composites, and suspension-based resins.

Sintering involves processing the component at high temperatures, resulting in a final 3D object composed of a single, fused material with low porosity. To reach this phase, the part must undergo a debinding process, where it is heated to a temperature that allows the binder to decompose and/or evaporate. This process is typically performed at low temperatures (between 200 °C and 300 °C) in the first stage and at high temperatures (between 300 °C and 600 °C) in the second stage [3]. At low temperatures, the binder begins to melt and expels gas, forming interconnected pores. At high temperatures, carbon oxidises due to binder decomposition, which releases expanded CO_2 and leads to the formation of cracks. Removing organic material at this stage can cause parts to malfunction, resulting in shrinkage and porosity. Therefore, the sintering process follows, which involves processing the part at higher temperatures to enable the compaction of the solid particles obtained from the debinding stage. The process can be broken down into three stages: first, the crystallites merge and recrystallise; second, there is an intermediate phase where particle adhesion begins and grain growth is observed; and finally, in the densification stage, the material fully compacts [4].

AM processes may vary depending on the type of 3D printer. However, a schematic representation of the general fabrication process is illustrated in figure 1.1.

While the process shown in figure 1.1 exemplifies the general overview of the 3D printing process, variations of it require proper clarification.

Driven by the demand for parts with high resolution, good mechanical and optical properties, ease of fabrication, and other factors, a variety of 3D printing methods have emerged over the years. These processes can be broadly categorised into five main groups: laser-based processes, extrusion processes, material jetting (MJ) processes, adhesive processes, and electron-beam-based processes.

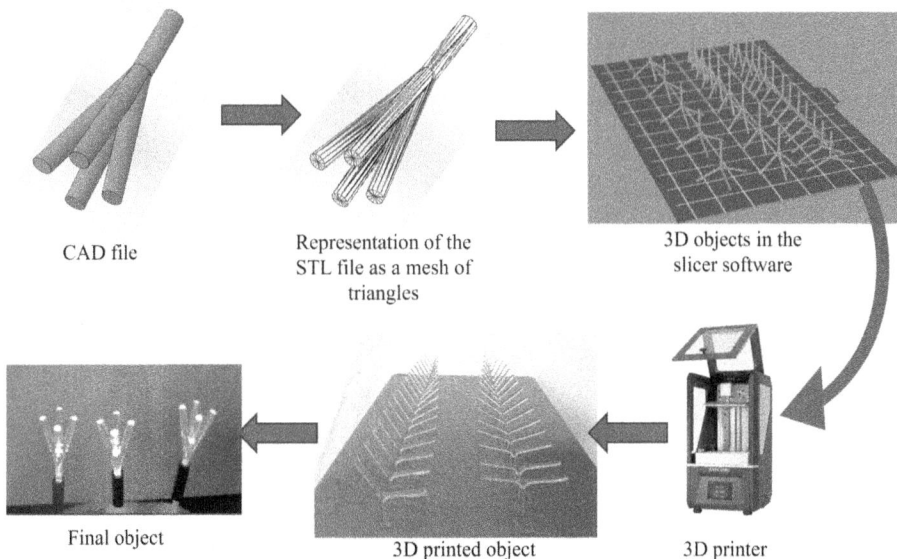

Figure 1.1. Additive manufacturing processes: CAD design, sliced model, 3D printing.

1.2 Stereolithography

One of the most widely used techniques in AM is stereolithography (SLA). SLA was invented by French and American scientists in the early 1980s and patented by American Charles W Hull in 1984 [5]. Since then, its use has become widespread, being nowadays a growing market [6]. It is predominantly used for professional and industrial applications, but the emergence of low-cost 3D printers has made the technology available to consumers and hobbyists. This technique is also known as the vat polymerisation technique, as the liquid resin is poured into a vat (or reservoir). In this technique, a UV light source tuned to the peak absorption wavelength of the photopolymerisable resins (commonly between 350 and 450 nm) selectively irradiates the liquid resin, leading to its photopolymerisation through cross-linking processes [7]. Most materials are based on pure photosensitive resins, but versions containing ceramic particles, metals, and composites can also be employed, depending on the specific requirements. The fabrication process is carried out using a layer-by-layer approach until the 3D object is complete.

1.2.1 Scanning-based lithography and projection-based lithography

Currently, various versions of SLA printing are available, and they have been categorised according to the light projection system and the growth direction. Figure 1.2 depicts schematic representations of different 3D printer configuration scenarios.

Regarding the light projection system, the two known SLA techniques are scanning-based stereolithography (SSL) [8] and projection-based stereolithography (PSL) [9]. In the SSL technique, each layer is created by scanning a collimated laser

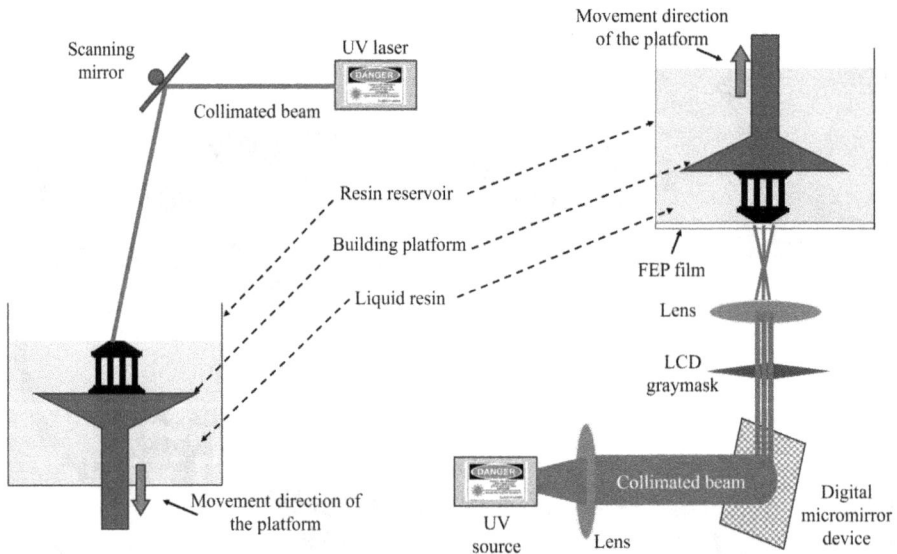

Figure 1.2. SLA 3D printing configurations categorised according to the illumination system: laser scanning (left side) and projection system (right side); and printing direction: bottom-up (left side) and top-down (right side). The printing direction and projection system are interchangeable.

beam with a predefined pattern, as shown on the left-hand side of figure 1.2. Laser tuning is achieved through stepper motors that move the laser beam or via a galvanometer that rotates a mirror, reflecting a stationary laser beam to the desired position. Later, the building platform moves away from the vat by a distance equivalent to one layer thickness, allowing the new space to be filled with liquid resin. The process is then repeated until the projected object is fully 3D printed.

In SSL, the laser beam must trace the entire sliced profile line by line. As a result, the fabrication time depends not only on the material's curing speed but also on the laser scanning velocity. Regarding resolution, in SSL, the beam diameter determines the details of the final print, necessitating small-diameter beams to achieve fine details in the printed parts. These drawbacks motivated the development of the second generation of SLA printers, specifically those based on masked stereo-lithography (MSLA) and digital light processing (DLP), also known as PSL.

In MSLA, the system utilises a liquid crystal display (LCD) mask [10, 11] to filter UV light from a wide-area light source. Therefore, the pixel size defines the resolution of the final print. Despite this, each layer is entirely processed at once, allowing faster speeds compared to SSL technology.

The DLP technique, as shown on the right-hand side of figure 1.2, works in the same fashion as MSLA, namely by processing a single layer at once. However, instead of an LCD filter to mask the light, it uses a digital micromirror device (DMD) [12] to cure the resin layer selectively. The DMD consists of a semiconductor chip with thousands of mirrors, each acting independently to reflect (or not reflect) the UV radiation. Each of these mirrors forms a pixel in the projected pattern. However, the resolution is determined by the dimensions of the pixels projected at the focal point. Thus, the DLP technology can achieve resolutions ranging from 10 to 100 μm.

Due to its ability to accommodate a large build area, SSL is well-suited for printing high-volume components [13]. On the other hand, PSL is more suited for low-volume components, as it can achieve higher resolutions using technologies such as LCDs and DMDs, which enable fine printing precision. Additionally, since PSL cures an entire layer at once, it often results in shorter printing times compared to SSL, which cures resin point by point.

Compared to other 3D printing techniques, such as fused deposition modelling (FDM), selective laser sintering (SLS), and selective laser melting (SLM), SLA can produce highly detailed features with resolutions typically ranging from 20 μm to 100 μm, though specialised systems can reach submicron resolutions under opti-mised conditions [12, 14]. Despite these advantages, SLA has certain limitations, including extended printing times due to the relatively low photopolymerisation rates inherent to the process. Furthermore, the geometric fidelity of SLA-printed components is limited by the layer-by-layer fabrication process, which can result in staircase effects on curved surfaces and abrupt transitions at layer boundaries. These limitations present challenges and thus remain an area of active research.

- **Printing growth**

Vat polymerisation printing can also be categorised according to the object growth direction, namely, from bottom to top and from top to bottom [15, 16], as shown in figure 1.2.

In the bottom-up approach, the surface of the building platform is parallel to the surface of the liquid photopolymerisable resin, and the top-to-bottom vertical motion of this platform relative to the surface of the liquid level defines the thickness of the 3D-printed object. For each position of the building plate relative to the surface of the resin (layer thickness), the UV radiation polymerises a predefined pattern. This allows the object to grow layer by layer from bottom to top.

In the top-down approach, the fabrication process begins with the building platform positioned near the bottom of the reservoir, which contains the liquid photopolymerisation resin. The gap formed between the bottom of the reservoir and the building platform is filled with liquid resin, corresponding to the thickness of the first layer of the object to be 3D printed. Next, the UV pattern exposes the resin from the bottom through a transparent fluorinated ethylene propylene (FEP) thin film that composes the bottom of the resin vat. The build platform then moves up slightly to allow the resin to fill the space between the cured resin layer and the bottom of the reservoir. The FEP film plays a crucial role in this step, as its non-stick properties prevent the 3D-printed part from adhering to the bottom of the container. The printing process is then repeated until the desired object is completed.

When comparing the bottom-up SLA approach to the top-down SLA approach, the latter is more advantageous as it requires less resin for printing, making it more cost-effective. Additionally, the top-down approach reduces the inhibition of the photopolymerisation process typically caused by oxygen exposure, as the resin surface is less susceptible to its effects. Resin refilling is also more efficient in the top-down method, as gravity helps spread the resin evenly over the build platform. In contrast, the bottom-up approach commonly relies on a roller to distribute the resin evenly across the platform and subsequent layers during the printing process.

1.2.2 Continuous liquid interface production

Continuous liquid interface production (CLIP) is considered the third generation of SLA 3D printing. In CLIP 3D printing technology, it is possible to fabricate objects continuously without the need for layer-by-layer construction, thereby mitigating the 'staircase effect'. As in top-down SLA 3D printing, the photosensitive resin is selectively photopolymerised through exposure to UV light. However, in CLIP, the transparent window that separates the resin reservoir from the UV source is composed of a membrane that allows oxygen to diffuse through. This inhibits the polymerisation of the resin in the region immediately above the membrane [17]. The thickness of this region depends on several parameters, including light intensity, oxygen concentration, build speed, and the viscosity of the resin [18]. The region is commonly known in the literature as the dead zone, and it allows the continuous upward movement of the building platform. This enables the continuous movement of the 3D-printed cured part, while the liquid resin is allowed to recirculate in the dead zone, thereby facilitating a continuous 3D printing process [19]. A representative illustration of the CLIP working principle is shown in figure 1.3.

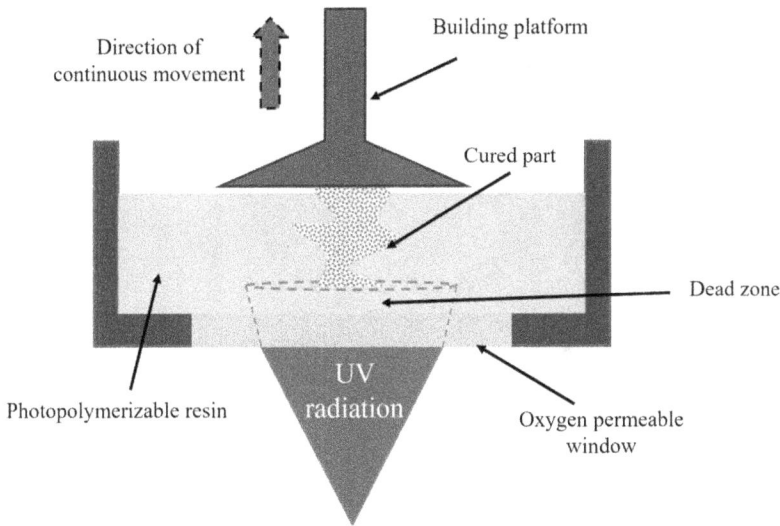

Figure 1.3. Representative schematic of CLIP. The oxygen-permeable window prevents the resin from polymerising right above the window, thus allowing continuous 3D printing.

The continuous production process in CLIP allows simultaneous UV curing and resin replenishment to take place without interruption. This enables a significantly faster printing process compared to its direct competitors, such as DLP and SLA. Additionally, continuous printing prevents the side effects associated with the stair-like appearance found in other vat polymerisation techniques. Despite its advantages, the CLIP technology also has downsides, such as the cost of the transparent membrane and the printing speed, which is a function of the viscosity and cure rate of the employed resin. Thus, resins with low viscosity are preferred over their counterparts, as the replenishment of the liquid resin in the dead zone must be fast enough to allow for quicker printing times.

1.3 Selective laser sintering and selective laser melting

SLS was developed and patented by Dr Carl Deckard and his academic adviser, Dr Joe Beaman, at the University of Texas at Austin in the mid-1980s [20]. SLS is an AM technique capable of producing components characterised by complex geometries without the requirement for support structures [21].

One of the features that makes SLS so compelling relates to its versatility in terms of material usage. This technique allows the production of 3D parts from different materials, including polymers, metals, ceramics, and composites [22], depending on the type and capabilities of the SLS machine.

SLS involves the use of a high-power laser, such as a CO_2 laser, to fuse a specific material powder into a desired three-dimensional shape. The first step in the SLS process involves preheating the bulk powder material located in the building chamber. This is done at a temperature below the material's melting point. Subsequently, the second step involves scanning a laser beam over the layer of

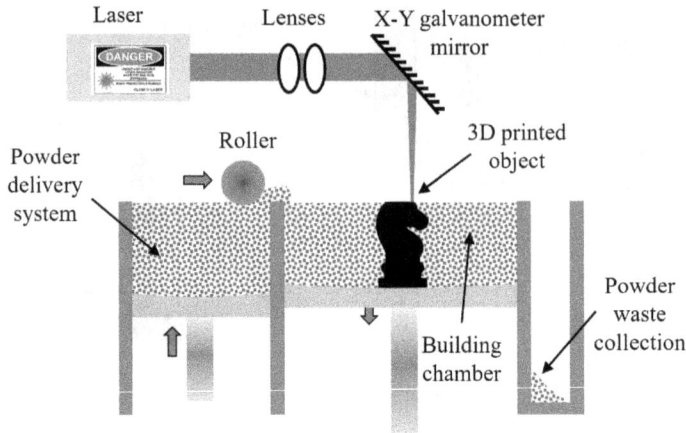

Figure 1.4. Working principle of SLS. From left to right: the roller pulls the powder material into the building chamber, and the laser is steered across the newly formed powder layer, allowing it to fuse. The process is repeated until the 3D-printed part is obtained.

powder material. This raises the temperature of the powder, enabling it to fuse. After the cooling of the irradiated regions, the first solid layer is created. To continue the process for the subsequent layers, the building platform descends by an increment equivalent to the layer's thickness, and a recoating roller fills the space with additional powder material. This cycle of layering and sintering is repeated until the desired 3D component is manufactured [23]. A representative schematic of the SLS process is shown in figure 1.4.

Compared to other AM processes, SLS can be considered more cost-efficient due to the absence of support structures. This occurs because the powder material can support the newly formed 3D parts without constraints related to gravitational force. Thus, the non-irradiated powder material can be reused for other prints, reducing the amount of waste material. Another key feature of this technique is the rigidity and resistance of the newly formed printing layers to mechanical stress. A major drawback of this process is that rapid cooling can cause deformation and shrinkage of the printed parts, which inherently affects the refractive index of 3D-printed optical components due to the strain optic effect, ultimately compromising performance. Additionally, poor surface finishing and dimensional inaccuracies are other significant limitations of this technique [24]. SLM employs the same concept as SLS; however, there are a few technical differences. One of the main differences between these processes is that in SLS, the powder is sintered or partially melted by the laser, whereas in SLM, the powder is completely melted. For this to happen, a high-intensity laser beam is used, providing enough energy to melt each 3D-printed layer. As the powder material melts, it enables the production of dense and compact parts, thereby avoiding the post-processing treatment required in the SLS technique.

Despite the opportunities presented by SLM, there are also some issues related to the quality of SLM parts, including possible overheating, keyholing, and incomplete fusion. This requires the optimisation of different parameters, such as the scanning strategy, the laser parameters, the support design, and the powder spreading method [25]. Among these, the most important are the laser parameters. This is because the laser is the functional unit that provides the necessary energy to melt the powder material [26] within a specific time frame that is controlled by the scanning speed. Another important parameter is the powder spreading method, as it is indispensable to have good uniformity of powder particle size in the SLM process [27].

1.4 Extrusion processes

FDM is a 3D printing method first developed in the 1990s by Stratasys, Inc. In this process, a thermoplastic filament is heated to its softening point by a heated liquefier. Examples of commonly used FDM filaments include acrylonitrile butadiene styrene (ABS), polylactic acid (PLA), polyethylene terephthalate glycol (PETG), polycarbonate (PC), and PC-ABS blends. Transparency is a key aspect when designing waveguides or diffusing optics. Thus, the filament choice should be made based on the final application.

During the fabrication process, the filament materials are extruded through a nozzle, forming a thin extruded filament. The diameter of this thin filament defines the resolution of the final 3D-printed part. The deposition is performed in the XY-plane using linear positioners, which can either move the printing head relative to the printing platform or vice versa, depending on the printer's design. This builds the first printed layer. For each of the subsequent layers, the printing head moves upward (or the building platform moves downward, depending on the printer model) by a distance equivalent to one layer thickness. The extrusion process is continuous, and the subsequent layers are formed on top of each other until the 3D object is constructed. A representation of this process is shown in figure 1.5.

In FDM printers, the temperature of the print bed or build platform can significantly impact the quality of the final 3D-printed component. There are two types of print bases: hot and cold. A cold base is one where no heat is applied to the platform, and the material is directly extruded onto the surface of the print base that is at room temperature. Conversely, in a hot print base, the temperature is controlled so that the temperature difference between the base and the material being extruded is minimal and certainly smaller than in the previous case. This reduced temperature differential helps to prevent issues associated with warping and cracking that can occur when the material cools too quickly. Such problems mainly occur in the first layers to be printed.

One important aspect to consider when using FDM printers is the need for support structures when printing free-form geometries with complex designs to prevent sagging and failure of the printed part. In dual-extruder printers, a second nozzle is often used to print support material, which is different from the main component's material. This allows easier removal after printing. However, in single-nozzle printers, supports are typically added using the same material as the 3D part and must be manually removed. The process of removing support structures is the first step in the post-processing of 3D-printed components. These support structures can be either soluble or standard. Standard supports are typically easy to remove,

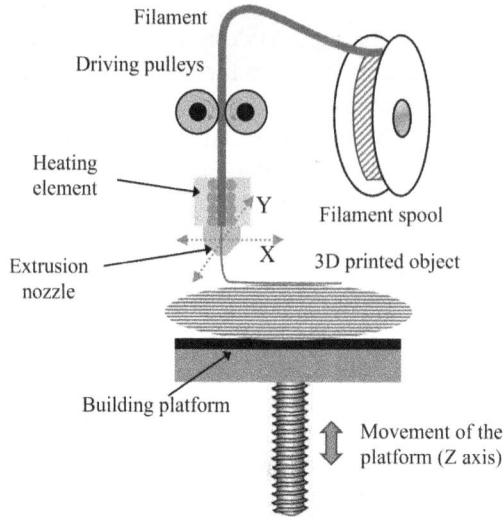

Figure 1.5. Schematic of the working principle of the FDM process. A thermoplastic filament is extruded layer by layer to create a 3D-printed object. The printing head moves in the XY-plane, and the build platform moves in the Z-axis.

except when they are placed in tight hollow regions in the component. In these cases, it is essential to consider the location of the support structures within the printed part prior to the fabrication process. On the other hand, water-soluble supports, such as poly(vinyl alcohol), are a possible alternative [28].

Once the support structures are removed, the next step typically involves treating the surface of the 3D object. This can be done through sanding processes [29], which smooth the surface and remove possible marks left by the supports and/or visible printing layers in detailed, curved parts. Furthermore, the surface of the 3D-printed object can be subjected to heat treatment, allowing it to reach a more refined state. Another effective method for enhancing surface roughness involves exposing the object to an acetone vapour bath, which significantly improves the surface quality of the component [30]. Additional methods, such as milling and drilling, can also be applied to soften the surface and assist with support removal. These post-processing steps are crucial for enhancing the surface finish of the component. Furthermore, they also ensure good adhesion when the component undergoes painting or coating processes.

Currently, intensive research is focused on improving the quality of the FDM 3D printing process. The quality of components produced through FDM depends on several parameters, including layer thickness, raster angle, part orientation, and air gap. Additionally, the number of layers can affect the temperature gradient. This gradient increases with the number of layers, which can have a positive influence on material diffusion. However, this can also lead to distortion between layers. The raster angle plays a significant role in the component's strength: a smaller raster angle results in longer raster paths, which increase strength but can also introduce more micro-stresses in the component. Finally, the air gap is crucial; reducing the air gap improves diffusion but may hinder heat transfer [31]. Research advancements in

FDM continue to enhance its precision and reliability, making it a promising solution for a wide range of applications in the future, including those dedicated to optics and photonics.

1.5 Material jetting

MJ is a photopolymer-based inkjet printing technology that, when compared to other AM technologies, exhibits promising features in terms of printing speed, resolution, and material selection. The printing process begins by heating the liquid resin to temperatures ranging from 30 °C to 60 °C, allowing it to reach an ideal level of viscosity for jetting. Then, tiny quantities of photopolymer are jetted onto the building platform, either continuously or using drop-on-demand methodologies, to produce the intended pattern. For this to happen, the printing head moves to the desired position above the building platform. The thin patterned layer of droplets is instantly photopolymerised by a UV source placed in the printing head [32], using wavelengths that correspond to the peak absorption of the resins, i.e. 190–400 nm [7]. After each layer is deposited and cured, the build platform moves downwards by a distance equivalent to the layer height, and the process is repeated in a continuous loop until the desired 3D component is formed. During the MJ process, a wax or gel-like support structure is also jetted to support the printed material, as the resin is typically in a liquid or molten state. The nozzle responsible for ejecting this gel-like material is also positioned in the printing head, close to the photopolymer printing nozzle. After the printing process, these support structures can be removed by high-pressure water jets or heating [33]. A representative diagram of the MJ process is shown in figure 1.6.

The levelling blade shown in figure 1.6 is used in MJ to control the layer thickness throughout the printing process. Its primary functions are to ensure each layer remains completely flat and to distribute the resin evenly. The levelling mechanism must also remove any excess material from the jetted droplets without compromising print quality. However, in some cases, a levelling blade may not be the most effective tool for this process. As a result, alternative methods, such as a roller module, may be used to obtain better material distribution and surface uniformity [34]. The minimum layer thickness of MJ printing is 16 μm [35], a very high resolution among AM printing technologies; it thus reduces the staircase effect produced by most AM technologies. Therefore, the 3D-printed parts exhibit low surface roughness, a highly desirable characteristic in optical and photonic applications. Furthermore, this can be further enhanced through two finishing options: a matte finish, achieved by covering the entire surface with the support material for a uniform texture, or a glossy finish, where exposed surfaces remain smooth and reflective while the support material is applied only where necessary [36].

1.6 Binder jetting

The inkjet-style printhead and the layer-by-layer approach used by the MJ method are characteristics shared by another AM technology, namely, the binder jetting (BJ) technique. However, the types of liquid-dispensed material and printing material are

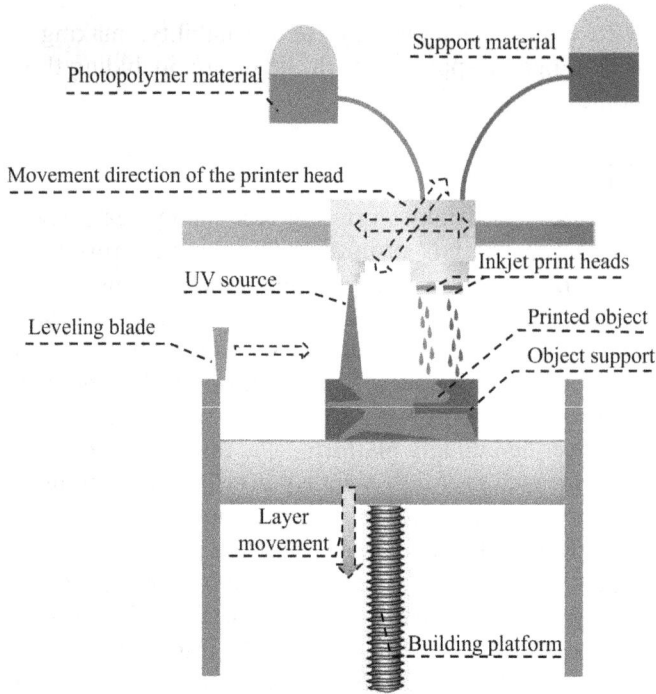

Figure 1.6. Representative schematic of an MJ printer. One nozzle is responsible for jetting the photopolymer resin, while the other is used for jetting the support material structure. The nozzles can move horizontally, while the build platform moves downwards every time a layer is completed.

very different, and this makes these techniques completely distinct in both manufacturing and post-processing.

BJ 3D printing utilises a two-chamber system, with one chamber for storing fresh powder and the other for constructing the part. A roller mechanism transfers powder from the storage chamber to the build chamber, simultaneously levelling the powder bed for uniform layer deposition. Subsequently, the binding agent is dispensed through an inkjet printhead onto the powder layer present in the build chamber. Liquid is dispensed over a predefined XY path generated from the sliced CAD model file. The reaction of the binding agent with the powder allows the powder to harden along the generated XY path. This completes one layer of the 3D object. The build platform is then lowered by a distance equivalent to the thickness of one layer, and the process is repeated for the subsequent layers until the designed model is completely 3D printed. A representative scheme of the binder jetting printer process can be seen in figure 1.7.

Once the printing process is complete, the freshly printed part still requires post-processing methodologies to be fully finished. The first step involves the curing process, which, depending on the binder agent, can be achieved through temperature or UV curing. Then, the unbound powder is removed, usually by vacuuming it into a receptacle where it can be recycled for future prints. Any remaining powder on the

a)

3D printed
object

Roller

Powder
waste
collection

Building
chamber

Feeding chamber

b)

Movement direction of the printer head

Printer head

Roller

Binder material

3D printed object

Powder
waste
collection

Building
chamber

Feeding chamber

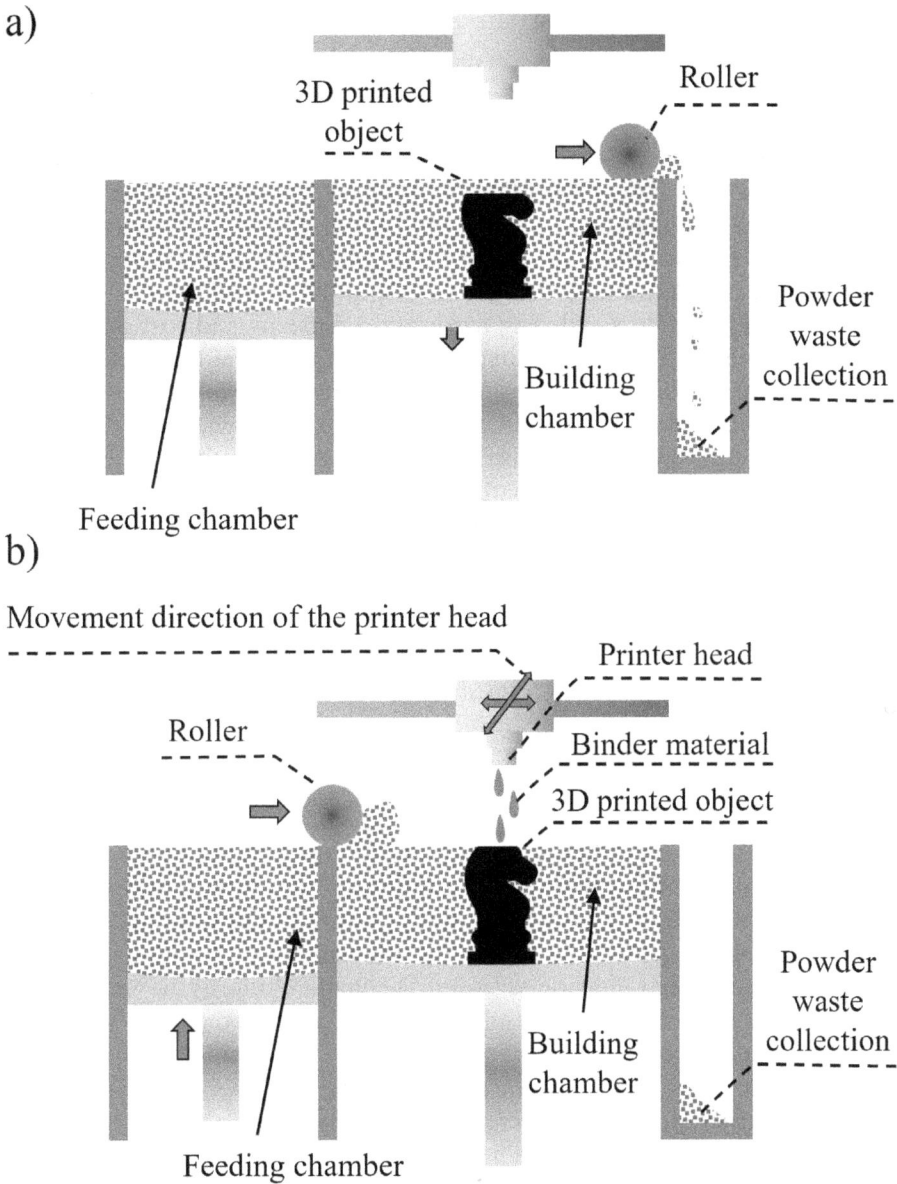

Figure 1.7. BJ printer steps: (a) the layer production step based on powder levelling and (b) the binder deposition step.

part is then cleaned off, typically using compressed air or a brush. An optional step used to enhance the mechanical properties of the printed component is infiltration with an additional binder. This infiltration strengthens the part, making it easier to handle without risking structural damage [37]. Depending on the material used, other optional post-processing steps may include debinding and sintering [38].

Compared to other AM technologies, such as SLS or SLM, the BJ process offers a speed advantage. This arises because the printhead only deposits the binder agent, whereas in other laser-based technologies, the sintering and melting processes occur during printing. This allows for a significant reduction in printing time, even considering the post-processing required in BJ. Furthermore, BJ printers are equipped with several printheads, allowing the binder to spread over a larger area along the print layers, thereby accelerating the printing process. Another advantage of BJ is that it is significantly more efficient and, consequently, less expensive than its competitors. This is because BJ printers use a printhead similar to those found in conventional inkjet printers, reducing hardware complexity and cost. Additionally, BJ operates under ambient room conditions, eliminating the need for specialised environments that require inert gases or high temperatures. This further reduces the overall cost and complexity of the process. Moreover, the unused powder from the printing process can be recycled, with only small amounts being wasted; thus, BJ minimises material waste and is considered environmentally friendly [39, 40]. Furthermore, BJ can utilise a wide range of powders, including metals, ceramics, and polymers, making it highly versatile compared to its competitors [37].

Despite all the advantages described above, the BJ process also has a few downsides. Among these are the laborious post-processing treatments required for the 3D part to achieve the desired mechanical properties, density, and strength, as well as the granular aspect of the printed parts, which is visible in the form of poor surface roughness.

1.7 Electron beam melting

Electron beam melting (EBM) is a relatively recent AM technology that has a working principle identical to that of the scanning electron microscope. In EBM, electrons are emitted from a tungsten filament and then collimated and accelerated to a kinetic energy of up to 60 keV. The electron beam is then passed through two electromagnetic coils that function as lenses. The first one adjusts the beam diameter to a spot size as small as 0.1 mm, while the second controls the beam's focal point within the building chamber. This entire system, known as an electron beam gun, remains fixed throughout the printing process [41].

Within the printing chamber, there is a raking system that forms the different powder layers. These can range from 5 to 200 μm. For printing to occur, two distinct powder-melting stages are required. Initially, the electron beam scans the powder layer at a high speed (10^4 mm s^{-1}). This preheats the material for the subsequent stage. Then, the beam is scanned at a lower velocity (10^3 mm s^{-1}), allowing the melting of the powder at desired locations and consolidating the material into a dense layer. After each printed layer is complete, the building platform moves down by a distance equivalent to the thickness of one layer. This process is repeated until the entire component is completed [42]. The fabrication is carried out in a vacuum environment, with typical vacuum pressures in the build chamber and electron beam gun of 10^{-3} and 10^{-5} Pa, respectively. A representative example of an EBM printer is shown in figure 1.8.

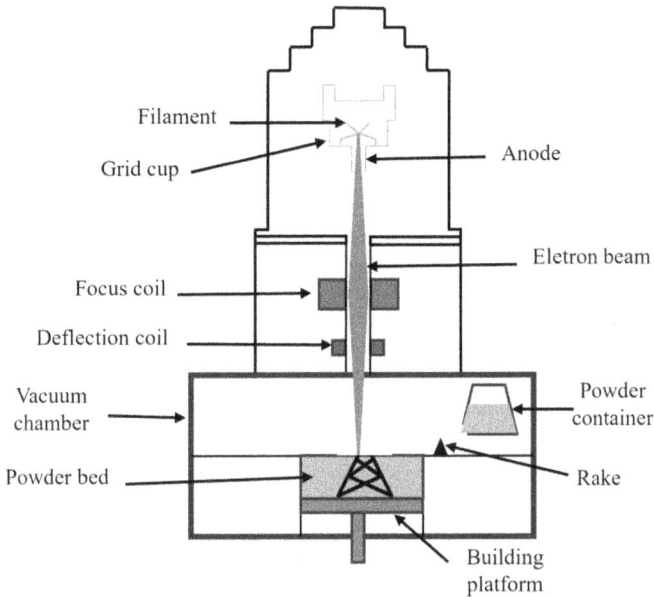

Figure 1.8. Representative schematic of an EBM printer. The top part operates similarly to an electron microscope, while the bottom part, i.e. the building chamber, is where the powder is melted layer by layer.

After the printing process is complete, the component cools down. This process can be hastened by adding helium gas to the chamber. Once the cooling is finished, the printed components are carefully removed from the EBM printer. Typically, these components are covered by a residual thin layer of metal powder that needs to be removed and reused for future print jobs, making this process environmentally friendly. Additionally, the remaining powder left in the printing chamber is also recycled for subsequent prints.

As an advantage, the EBM method enables the production of components with high purity, resulting in printed parts with low concentrations of impurities, such as oxygen and nitrogen. This is particularly valuable to industries such as aerospace and healthcare, where material purity is critical. This high purity is achieved using a vacuum chamber that removes most impurities present in the chamber during the printing process. Additionally, EBM allows the fabrication of parts with low porosity. The research literature indicates that the highest porosity typically achieved is less than 200 μm. This low level of porosity reduces the need to use complex post-processing treatments. Furthermore, this method allows the creation of a homogeneous microstructure and a smooth surface finish. Finally, the electron beam can melt almost any metal or alloy, making this method particularly suitable for producing free-form optics.

1.8 Laminated object manufacturing

Laminated object manufacturing (LOM) is an AM technology introduced in the 1990s. This process was initially developed for rapid prototyping and was later adapted for other applications, including the manufacture of functional parts. Unlike standard powder- and liquid-based AM methods, LOM employs sheets of material that are stacked, bonded, and cut to form the desired shape.

The working principle of the LOM process consists of four steps: layering, bonding, cutting, and post-processing. The process starts with the preparation of the building platform. For this, a thin sheet of adhesive-coated material, such as paper, plastic, or metal, is fed onto the building platform using a feed roller. After this, a heated roller evenly presses the sheet against the platform, activating the adhesive and ensuring a strong bond between the layers. Then, the layer-cutting process takes place. Here, the layer is cut to the desired shape, following the positioning given by the sliced file, using a knife or a high-power CO_2 laser. The latter is more advantageous due to its precision, efficiency, and ability to process various materials. Despite this, it is necessary to calibrate the system so that each cutting tool cuts just one adhesive layer at a time. This terminates the first layer fabrication. For the subsequent layers, a new adhesive-coated layer is applied on top of the previously printed layer, bonding them together. Then, the cutting and layering processes repeat in a continuous cycle until the desired 3D-printed part is completed.

After 3D printing, it is necessary to remove the excess sections of unused material from the component. To facilitate this process, cross-hatching is always added to the excess material during the cutting process [43]. This ensures that the leftovers can be easily removed (manually) at the end of the fabrication process. A representative schematic of a LOM 3D printer is shown in figure 1.9.

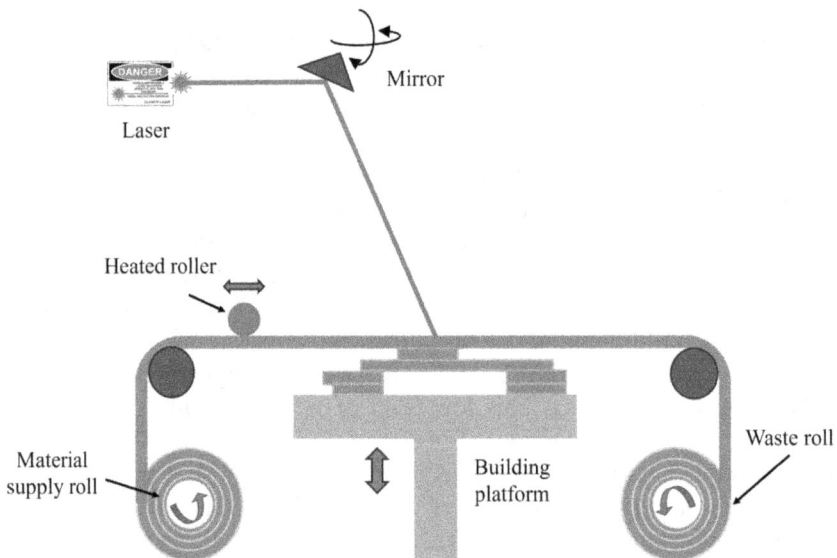

Figure 1.9. Representative schematic of an LOM 3D printer. A thin adhesive sheet of material is fed through a feed roller. A heated roller presses the sheet against the previous layer of the 3D object, ensuring adhesion. A laser is used to cut the layer to the desired shape, and the process is repeated.

LOM offers several advantages compared to other AM processes. With LOM, it is possible to process materials such as paper, polymers, ceramics, and composites, unlike many AM technologies that rely on resins and powders. Furthermore, LOM also offers the possibility of using strands of paper and strands of polymer to produce parts with complex 3D geometries. This is advantageous compared to other technologies, as it uses nontoxic and inexpensive materials to produce the components [44, 45]. Additionally, the printed parts are known to have low internal stress, reducing the risk of shrinkage, deformation, or distortion. Furthermore, LOM-printed objects exhibit lower brittleness and fragility, resulting in enhanced resistance. Finally, one of the most significant advantages of the LOM method is its capability to produce large components, thereby overcoming one of the primary limitations of several AM methods.

The drawbacks of this technology are associated with the decubing process, which can sometimes be labour-intensive and time-consuming, depending on the intricate features produced on the 3D printed part. It is also important to note that certain materials may expand during the printing process. This causes the layers to become uneven, affecting the final shape of the printed part. Additionally, low-quality cuts result in low-detail components. Therefore, it can be concluded that LOM is suitable for the manufacture of large components that do not require detailed designs.

1.9 Two-photon polymerisation 3D printing

Nowadays, the demand for components at the micro- and nanoscale is increasing due to advancements in optical devices for photonic integrated circuits, optical communications and data processing, high-resolution microscopy, and sensing, among others. This has led to the necessity of finding methods that allow the production of high-resolution components. In this respect, AM is no exception, so it is no surprise that AM technologies have already been proposed to produce components at the micro- and nanoscale. Direct laser writing technologies are among the proposed methods capable of building three-dimensional components at the nanoscale with high accuracy and precise dimensions.

Among the laser-based methods, the two-photon polymerisation (TPP) technology, also known as two-photon lithography (TPL), is gaining momentum in this field [46]. The reason for this is related to its ability to manufacture objects with resolutions smaller than the optical diffraction limit of the laser used [47].

In a typical TPP printer, a collimated femtosecond (fs) laser beam operating in the visible or near-infrared region is passed through an attenuator and a beam expander to control both the intensity and diameter of the laser beam, respectively. The beam expansion is required to overfill the back aperture of the objective, enabling it to fully utilise its numerical aperture. The beam then passes through a shutter, such as an acousto-optic modulator controller, which modulates the laser transmission, allowing control over the exposure time. The beam is later reflected in a pair of galvo mirrors, which are responsible for the movements defined by the sliced file. The beam is further passed through a series of mirrors that guide it to the objective lens, with the latter being responsible for focusing the beam onto the photoresist. The system is also equipped with a dichroic mirror and a CCD camera to facilitate the alignment process and enable real-time visualisation. A representation of a typical TPP printer is shown in figure 1.10.

TPP requires materials to be transparent to the fs laser wavelength, allowing the radiation to penetrate without loss and enhancing the energy absorption efficiency. Furthermore, the polymerisation threshold needs to be lower than the ablation threshold to prevent degradation or material loss. Finally, a solvent capable of washing the 3D-printed parts is necessary.

The key components of the materials used for TPP printing consist of a mixture of oligomers and monomers, as well as the photoinitiator. The first corresponds to the building material that will make up the final 3D-printed part, while the second is responsible for the absorption of the two photons of incident light, triggering the photopolymerisation reaction in the focused area. Depending on the specific printing requirements, the ratio of these two components may vary.

In TPP, both negative and positive photoresists can be used. When a negative photoresist is used, two-photon absorption cures the material in the exposed regions, while the uncured parts are washed away through simple methods. When a positive photoresist is used, polymer chains in the irradiated areas are broken into small fragments, which then require dissolution and removal through additional steps to preserve the unexposed areas. This makes post-processing more complex; therefore, negative photoresists have been the preferred choice.

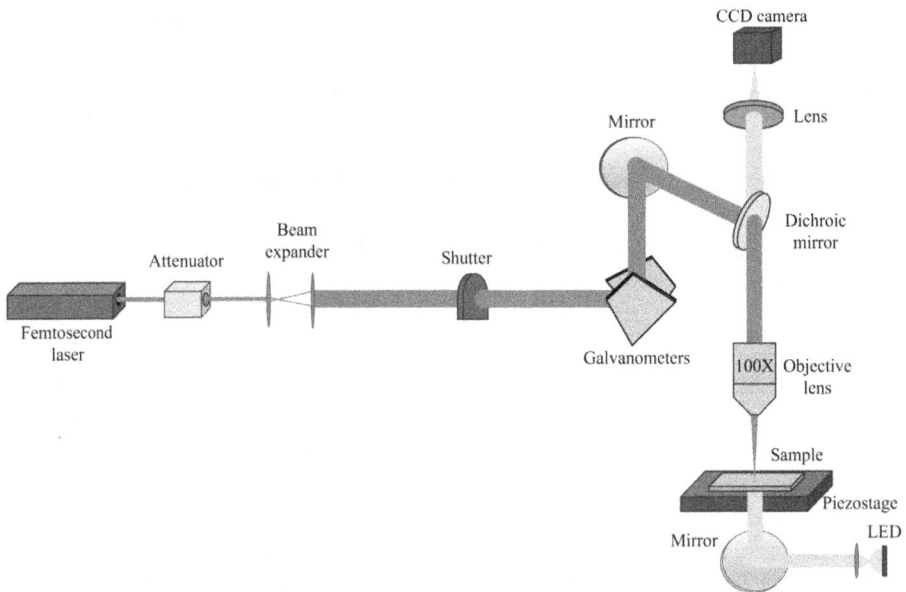

Figure 1.10. Representative schematic of a TPP 3D printer. A femtosecond laser is focused and scanned in a specific pattern over a drop of liquid photoresist.

Through the correct selection of material and slice parameters, such as exposure, velocity, printing orientation, and voxel size (the latter of which is determined by the objective lens), it is possible to develop 3D parts at the micro- and nanoscale. Working at these scales and using transparent materials makes it possible to manipulate light radiation with unprecedented precision. Thus, TPP is poised to revolutionise the photonics world.

References

[1] Piedra-Cascón W, Krishnamurthy V R, Att W and Revilla-León M 2021 3D printing parameters, supporting structures, slicing, and post-processing procedures of vat-polymerization additive manufacturing technologies: a narrative review *J. Dent.* **109** 103630

[2] Riccio C *et al* 2021 Effects of curing on photosensitive resins in SLA additive manufacturing *Appl. Mech.* **2** 942–55

[3] Wang K *et al* 2020 Study on defect-free debinding green body of ceramic formed by DLP technology *Ceram. Int.* **46** 2438–46

[4] Kuang X, Carotenuto G and Nicolais L 1997 A review of ceramic sintering and suggestions on reducing sintering temperatures *Adv. Perform. Mater.* **4** 257–74

[5] Hull C W 1984 *Apparatus for production of three dimensional objects by stereolithography* US US-4575330-A

[6] Sterolithography SLA Technology 3D Printing Market https://knowledge-sourcing.com/report/sterolithography-sla-technology-3d-printing-market (accessed 12 October 2023)

[7] Bagheri A and Jin J 2019 Photopolymerization in 3D printing *ACS Appl. Polym. Mater.* **1** 593–611

[8] Jacobs P F 1992 *Rapid prototyping and manufacturing: fundamentals of stereolithography* (Dearborn, MI: Society of Manufacturing Engineers)

[9] Emami M M, Barazandeh F and Yaghmaie F 2015 An analytical model for scanning-projection based stereolithography *J. Mater. Process. Technol.* **219** 17–27

[10] Pomerantz I *et al* 1991 *Three dimensional modeling apparatus* US US-5031120-A

[11] Bertsch A, Zissi S, Jézéquel J Y, Corbel S and André J C 1997 Microstereophotolithography using a liquid crystal display as dynamic mask-generator *Microsyst. Technol.* **3** 42–7

[12] Sun C, Fang N, Wu D M and Zhang X 2005 Projection micro-stereolithography using digital micro-mirror dynamic mask *Sensors Actuators* A **121** 113–20

[13] Cheung L K, Wong M C M and Wong L L S 2001 The applications of stereolithography in facial reconstructive surgery *Proc. of the Int. Workshop on Medical Imaging and Augmented Reality* (Los Alamitos, CA: IEEE Computer Society) pp. 10–5

[14] Zheng X *et al* 2012 Design and optimization of a light-emitting diode projection micro-stereolithography three-dimensional manufacturing system *Rev. Sci. Instrum.* **83** 125001

[15] Manapat J Z, Chen Q, Ye P and Advincula R C 2017 *3D Printing of Polymer Nanocomposites via Stereolithography* (Weinheim: Wiley-VCH)

[16] Melchels F P W, Feijen J and Grijpma D W 2010 A review on stereolithography and its applications in biomedical engineering *Biomaterials* **31** 6121–30

[17] Januszewicz R, Tumbleston J R, Quintanilla A L, Mecham S J and Desimone J M 2016 Layerless fabrication with continuous liquid interface production *Proc. Natl Acad. Sci.* **113** 11703–8

[18] Tumbleston J R *et al* 2015 Continuous liquid interface production of 3D objects *Science* **347** 1349–52

[19] Balli J, Kumpaty S and Anewenter V 2017 Continuous liquid interface production of 3D objects: an unconventional technology and its challenges and opportunities *ASME 2017 Int. Mechanical Engineering Congress and Exposition*

[20] Deckard C R 1987 *Method and Apparatus for Producing Parts by Selective Sintering* World International Property Organization WO1988002677A2

[21] Berry E *et al* 1997 Preliminary experience with medical applications of rapid prototyping by selective laser sintering computer visualization techniques for rendering *Med. Eng. Phys.* **19** 90–6

[22] Kruth J P, Wang X, Laoui T and Froyen L 2003 Lasers and materials in selective laser sintering *Assem. Autom.* **23** 357–71

[23] Olakanmi E O 2013 Selective laser sintering/melting (SLS/SLM) of pure Al, Al–Mg, and Al–Si powders: effect of processing conditions and powder properties *J. Mater. Process. Technol.* **213** 1387–405

[24] Wong K V and Hernandez A 2012 A review of additive manufacturing *ISRN Mech. Eng.* **2012** 1–10

[25] Yadroitsev I, Yadroitsava I, Bertrand P and Smurov I 2012 Factor analysis of selective laser melting process parameters and geometrical characteristics of synthesized single tracks *Rapid. Prototyp. J.* **18** 201–8

[26] Yasa E, Deckers J and Kruth J P 2011 The investigation of the influence of laser re-melting on density, surface quality and microstructure of selective laser melting parts *Rapid. Prototyp. J.* **17** 312–27

[27] Marcu T, Todea M, Gligor I, Berce P and Popa C 2012 Effect of surface conditioning on the flowability of Ti6Al7Nb powder for selective laser melting applications *Appl. Surf. Sci.* **258** 3276–82

[28] Ni F, Wang G and Zhao H 2017 Fabrication of water-soluble poly (vinyl alcohol)-based composites with improved thermal behavior for potential three-dimensional printing application *J. Appl. Polym. Sci.* **134** 44966

[29] Ahn S H, Lee C S and Jeong W 2004 Development of translucent FDM parts by post-processing *Rapid. Prototyp. J.* **10** 218–24

[30] Lalehpour A and Barari A 2016 Post processing for fused deposition modeling parts with acetone vapour bath *IFAC-PapersOnLine* (Amsterdam: Elsevier B.V.) pp. 42–8

[31] Sood A K, Ohdar R K and Mahapatra S S 2010 Parametric appraisal of mechanical property of fused deposition modelling processed parts *Mater. Des.* **31** 287–95

[32] Tyagi S, Yadav A and Deshmukh S 2021 Review on mechanical characterization of 3D printed parts created using material jetting process *Mater. Today Proc.* **51** 1012–6

[33] O'Neill P, Jolivet L, Kent N J and Brabazon D 2017 Physical integrity of 3D printed parts for use as embossing tools *Adv. Mater. Process. Technol.* **3** 308–17

[34] Cheng Y L, Chang C H and Kuo C 2020 Experimental study on leveling mechanism for material-jetting-type color 3D printing *Rapid. Prototyp. J.* **26** 11–20

[35] Sireesha M, Lee J, Kranthi Kiran A S, Babu V J, Kee B B T and Ramakrishna S 2018 A review on additive manufacturing and its way into the oil and gas industry *RSC Adv.* **8** 22460–8

[36] Gülcan O, Günaydın K and Tamer A 2021 The state of the art of material jetting—a critical review *Polymers (Basel)* **13** 2829

[37] Mostafaei A *et al* 2021 Binder jet 3D printing—process parameters, materials, properties, modeling, and challenges *Prog. Mater Sci.* **119** 100707

[38] Wang Y and Zhao Y F 2017 Investigation of sintering shrinkage in binder jetting additive manufacturing process *Procedia Manufacturing* (Amsterdam: Elsevier B.V) pp. 779–90

[39] Dini F, Ghaffari S A, Jafar J, Hamidreza R and Marjan S 2020 A review of binder jet process parameters; powder, binder, printing and sintering condition *Met. Powder Rep.* **75** 95–100

[40] Zhang Y, Jarosinski W, Jung Y G and Zhang J 2018 Additive manufacturing processes and equipment *Additive Manufacturing: Materials, Processes, Quantifications and Applications* (Amsterdam: Elsevier) pp. 39–51

[41] Biamino S *et al* 2011 Electron beam melting of Ti–48Al–2Cr–2Nb alloy: microstructure and mechanical properties investigation *Intermetallics (Barking)* **19** 776–81

[42] Sonkamble V and Phafat N 2023 A current review on electron beam assisted additive manufacturing technology: recent trends and advances in materials design *Discov. Mech. Eng.* **2** 1

[43] Liao Y S and Chiu Y Y 2001 Adaptive crosshatch approach for the laminated object manufacturing (LOM) process *Int. J. Prod. Res.* **39** 3479–90

[44] Mueller B and Kochan D 1999 Laminated object manufacturing for rapid tooling and patternmaking in foundry industry *Comput. Ind.* **39** 47–53

[45] Dermeik B and Travitzky N 2020 Laminated object manufacturing of ceramic-based materials *Adv. Eng. Mater.* **22** 2000256

[46] Maruo S and Kawata S 1997 Two-photon-absorbed photopolymerization for three-dimensional microfabrication *Proc. IEEE the 10th Annual Int. Workshop on Micro Electro Mechanical Systems. An Investigation of Micro Structures, Sensors, Actuators, Machines and Robots* 169–74

[47] Kawata S, Sun H-B, Tanaka T and Takada K 2001 Finer features for functional micro-devices *Nature* **412** 697–8

IOP Publishing

Additive Manufacturing in Optics and Photonics
Fabrication and applications
Ricardo Oliveira and Nuno Valente

Chapter 2

Three-dimensional printing of optical components

Optical components are generally expensive due to tight tolerances, labour-intensive techniques, and the limited concurrency of the market. Additive manufacturing (AM) is a technology that can operate concurrently, as it produces components with high resolution, specifically at the micrometre or even nanometre scale, using simpler methodologies in a cost-effective manner. Optical parts, traditionally produced through milling and grinding, can now be fabricated in a single step using readily available AM printers, streamlining the process and reducing costs. Nowadays, AM can 3D print parts at various length scales, ranging from nanometres to tens of centimetres, while maintaining high resolution, speed, and low cost. As a result, AM plays a more prominent role in fabricating optical components. Furthermore, AM printers can create complex geometries, including nonplanar surfaces, curvilinear substrates, and intricate 3D patterns. This design freedom signifies a new era in optical components, opening numerous possibilities for developing components that were previously difficult, costly, or impossible to manufacture using conventional production techniques. This development enables researchers and industries to produce and test their products in-house, reducing dependence on external suppliers and breaking the time-dependent loop often present during product conceptualisation and development.

Given the numerous opportunities that AM offers for developing optical components, several studies have already been conducted. Examples include the development of traditional and freeform lenses, mirrors, integrated spheres, and kinetic mounts, among other optical components that can assist in the development of optical components. This chapter will explore recent advancements in research that utilises AM in the manufacture of optical components.

doi:10.1088/978-0-7503-6428-7ch2   2-1   © IOP Publishing Ltd 2025. All rights,

2.1 Optical lenses

Lenses are optical components used in our everyday life. Typical examples include eyeglasses, mobile phone cameras, general-purpose cameras, light detection and ranging (LIDAR) systems, autonomous vision systems, a wide range of optical telecom components, microscopes, telescopes, and other devices. The typical production process of a lens consists of three main steps: grinding, polishing, and lapping [1, 2]. These can be considered laborious and time-consuming, thus restricting the use of complex geometrical shapes. As a result, AM technologies have been proposed. For this, the use of a 3D printer, suitable material, and a computer-aided design (CAD) model of the lens are generally the three fundamental components required to create the part. This reduces complexity, fabrication steps, and cost. Computer numerical control (CNC) machining enables the creation of intricate, freeform shapes; however, the high cost and time involved restrict the development of parts for mass production. Commonly used fabrication methods, such as injection moulding, also allow the fabrication of complex shapes. However, this technology requires the development of expensive moulds (i.e. through CNC) and is unsuitable for low-volume production.

This section will focus on the various types of lenses already produced through AM and the methodologies employed in their production. It will also explore the 3D printing materials most commonly used to make lenses and the results achieved.

2.1.1 Manufacture of lenses from the macroscale to the mesoscale

Optical lenses are the fundamental components of various optical systems, including eyeglasses, microscopes, telescopes, cameras, and optical communication devices. Thus, they are vital to the society in which we live. Lenses work through refraction, allowing the manipulation of light to perform functions such as focusing, collimating, magnifying objects, and correcting aberrations, among many other functions. The simplest type of lens, called a singlet, is the fundamental building block of complex optical systems. This lens type comprises a single material with one or two curved spherical surfaces. Its simplistic shape has inherent disadvantages; one of the most well known is related to spherical aberration, where light rays do not converge to the same focal point, resulting in blurred images. Chromatic aberration, astigmatism, and other issues are potential problems associated with singlet lenses. While multi-lens systems can solve this problem, they require the stacking of multiple lenses, which restricts their use in compact systems. Compact designs, such as the use of aspherical lenses to correct, for instance, spherical aberration, are one possible solution, as exemplified in figure 2.1.

However, manufacturing such lenses requires a complex machining process, which increases the cost of the lenses. The use of freeform lenses presents interesting opportunities for manipulating light. These lenses are characterised by their complex, non-symmetric surfaces, allowing them to perform special functions that would otherwise require a set of singlet lenses. Examples include the correction of various optical aberrations, such as spherical aberration, coma, and astigmatism, which leads to sharper images. Additionally, these lenses offer compact and

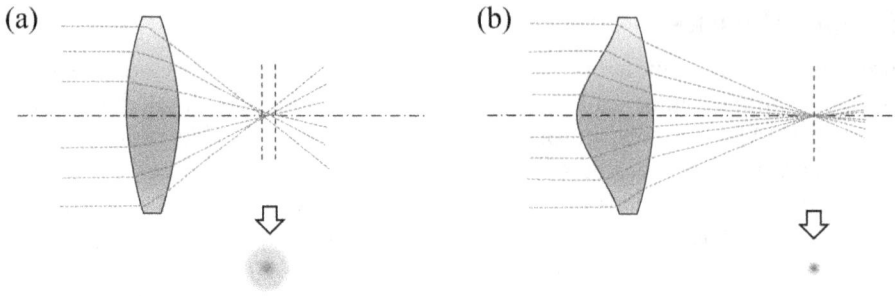

Figure 2.1. (a) Spherical aberration in a singlet double-convex spherical lens, and (b) its correction using an aspheric lens. The bottom images correspond to the spot-size images at the focal point.

lightweight designs, customised light distribution, precise beam shaping, and the integration of multiple optical functions. Again, the fabrication of such lenses through traditional fabrication techniques is labour-intensive, raising their cost and making them prohibitive to mass-produce.

Despite being well-established, lenses are precise optical elements that require fine-tuning of their materials and surface finishes. Thus, they rely on materials with high mechanical robustness and good optical performance while still requiring precise machinery and technologies to shape them. Overall, the surface precision of a lens needs to be on a scale of $\lambda/2$ (or worse if used in low-precision optics, as is the case with general imaging); $\lambda/4$ or better for high-quality optical components; and $\lambda/10$ or higher for precision optics. Considering an optical lens operating at visible wavelengths, such as 400 nm, and applying the $\lambda/4$ criterion, this requires a surface roughness of approximately 100 nm for a lens to be considered suitable for precision optics, which highlights the high level of accuracy needed and the potential challenges inherent in achieving such surface smoothness.

Some well-established methods for the precise production of lenses are polishing and grinding. For the mass production of lenses at a lower cost, techniques such as injection moulding or compression moulding [3, 4] are standardised. Using these technologies enables the manufacture of components with nanoscale surface quality with relative ease. However, when the geometry of the desired components is complex, containing freeform shapes and intricate details, these technologies face some challenges, while the alternatives are laborious and costly. Moulding-based methods allow the mass production of simple structures, such as standard aspherical or spherical lenses. Again, when complex geometries are required, they tend to be expensive due to the inherent difficulties in producing the mould [5]. Magnetorheological finishing is another method that produces high-quality precision surface optics. However, this method has limitations too, such as the difficulty of scaling production [6]. Diamond turning is another alternative, allowing the manufacture of optical components without post-processing. Conversely, it has disadvantages, such as the fact that some of the manufactured components present high scattering [7]. Considering all these constraints, the scientific community has already taken advantage of AM manufacturing to produce lenses.

One example of lenses fabricated through AM technologies was demonstrated by Gawedzinski *et al* [8], who compared plano-convex (PC) and biconvex lenses fabricated by the inkjet printing technique from Luxexcel with commercial glass lenses processed through traditional grinding and polishing methods. Some of the lenses produced in this work are shown in figure 2.2.

The results were promising, showing that the 3D-printed lenses had a surface roughness identical to that found in glass lenses manufactured by injection moulding, with an average roughness profile (Ra) of less than 20 nm. Furthermore, the root mean square (RMS) wavefront error was at least 18.8 times larger than that of the equivalent glass prototypes for a lens with a 12.7 mm clear aperture. However, when measured within 63% of its clear aperture, the RMS wavefront error of the 3D-printed lenses was comparable to that of glass lenses. Another example of the fabrication and characterisation of 3D-printed lenses through material jetting (MJ) was reported by Assefa *et al* [9], who demonstrated the manufacture of large-scale PC lenses with a diameter of 103 mm. The authors used a modified inkjet printing technology and employed an iterative error-correction process. The refined technique was also explored in other papers [10, 11], allowing the manufacture of components without post-processing treatments. The manufactured PC lenses demonstrated promising results, with a surface profile deviation of ±500 nm within a 12 mm aperture diameter. Their RMS surface roughness was less than 1 nm without surface polishing. These interesting results pave the way for diffraction-limited performance.

The additive manufacture of lenses has also been achieved using stereolithography (SLA) [12]. For this, the authors used the Form 2 printer from Formlabs, along with Clear Resin from the same manufacturer. The printing tests were conducted for PC aspheric lenses 12.7 mm in diameter, with parameters similar to those of traditional commercial glass lenses, namely LA1540 and LA1560, with focal lengths of 15 and 25 mm, respectively, and normal radii of curvature of 7.7 and 12.9 mm, respectively. The desired lenses were printed with exact dimensions and

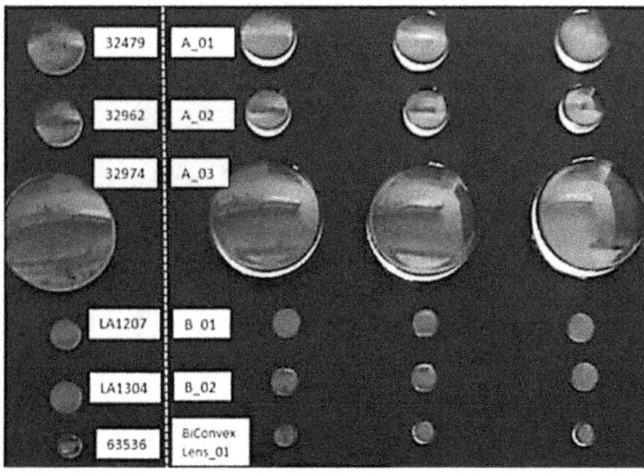

Figure 2.2. Commercial glass lenses (left) and 3D-printed Luxexcel® lenses (right). Reproduced with permission from [8]. © 2017 Society of Photo-Optical Instrumentation Engineers (SPIE).

characteristics and were later post-processed through spin coating or by directly curing the lenses onto glass concave lenses. These processes were crucial in eliminating the 'staircase' effect associated with the layered structure inherent to the printing process. Furthermore, the lens printing angle has also been considered to minimise the layer line effect and deformation. The printed results are shown in figure 2.3.

The characterisation results for the SLA 3D printed PC lenses presented in [12] demonstrate that the printing angle is crucial for achieving optimal performance. Printing lenses so that their midplane lies parallel to the XY plane of the printer results in the appearance of prominent layer lines, which, despite preserving the desired lens surface, inhibit the coating of the convex lens surface. On the other hand, printing lenses perpendicular to the printing bed results in lenses with more significant deformations and correspondingly worse form measurements. Results have shown that the best performance is achieved when lenses are printed at an angle of 60°. Yet, the placement of support structures needs to be considered at the border of the lenses, and this is also seen to have side effects on the lenses' final shape. Additionally, characterisation has revealed the presence of astigmatism and average RMS roughness values ranging from 13 to 28 nm, as well as RMS wavefront deviations between 0.297 and 0.374 waves for spin-coated lenses. This is acceptable for low-precision optical applications but still suboptimal for high-precision optics. However, the R_a RMS for the glass-cured lenses was 6 nm, and the average form RMS was 0.048 waves. Despite the need for further studies, this research revealed that consumer-grade 3D printers could present an opportunity to produce freeform lenses.

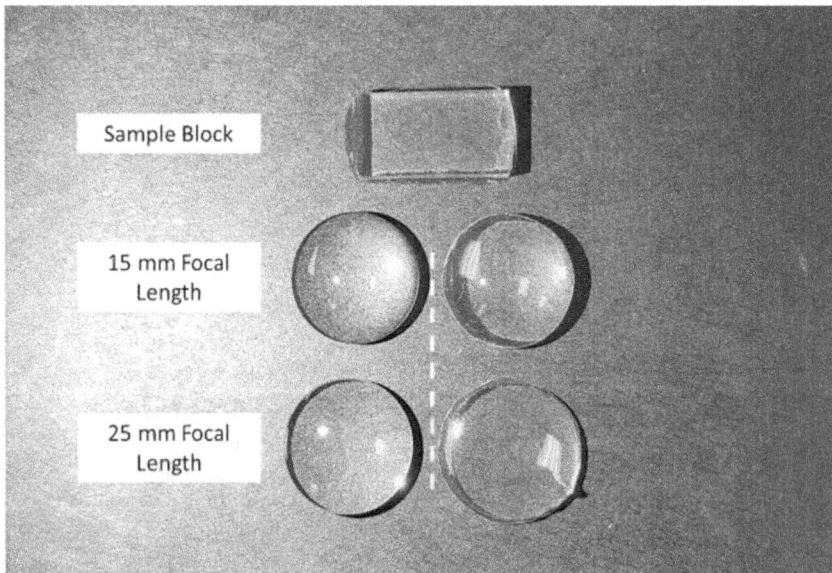

Figure 2.3. SLA 3D-printed PC lenses (right) vs. commercial glass lenses (left). Reprinted with permission from [12]. © 2019 Optical Society of America.

Typically, when manufacturing optical components, two key parameters are considered: dimensional accuracy and surface smoothness of the printed component. One technology that meets these two requirements is two-photon polymerisation (TPP), which has sufficient resolution to print microlenses. However, this technology is generally considered time-consuming, making it undesirable for large-scale manufacturing [13, 14]. The technological problems become more problematic when considering the stitching marks. These marks appear when larger, continuous 3D structures need to be printed. Since the scanning area is limited, it is necessary to combine multiple scanning areas, resulting in discontinuities between the scanning regions. There are already some advanced techniques based on the synchronisation of linear and galvanometric scanners for efficient femtosecond 3D optical printing of objects at the mesoscale, allowing for reduced stitching marks [15]. On the topic of optical lenses, Ristok and co-workers [16] demonstrated the 3D printing of spherical and aspherical lenses with diameters of up to 2 mm using a commercial TPP printer made by Nanoscribe GmbH. To verify the performance of the printed lenses compared to commercial lenses, the printed ones were evaluated against commercial glass lenses. The results for the 3D-printed lens and its corresponding glass counterpart are presented in figure 2.4.

Regarding imaging quality, the spherical lenses were identical to the commercial ones. However, the printed lens revealed a slight deviation in its optical design (see figures 2.4(b) and (c)). This slight deviation was associated with the shrinkage of the photoresist used. Nonetheless, the modulation transfer function of the printed lens was identical to that of the glass one (see figure 2.4(d)).

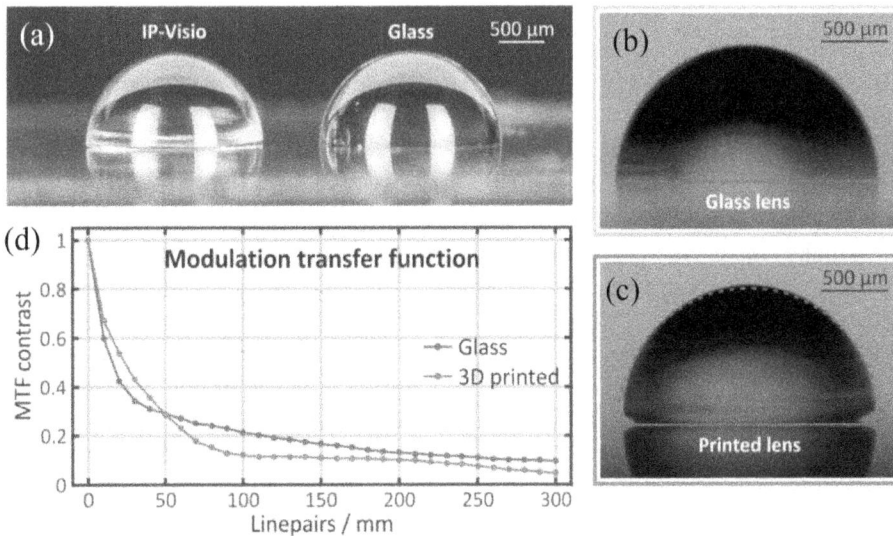

Figure 2.4. (a) Comparison of a 3D-printed lens and a commercial glass lens (2 mm diameter): TPP lens with reduced stitching errors (left) and glass lens (right). Lateral view of (b) the glass lens and (c) the TPP lens. (d) Modulation transfer functions of the glass and TPP lenses, considering normal incidence. Reprinted with permission from [16]. © 2020 Optical Society of America.

Using AM to fabricate optical components offers tremendous simplicity and cost-reduction opportunities. However, the layer-by-layer process inherent in most standardised SLA technologies or the line-by-line scanning used in laser technologies inevitably leads to time-consuming processes that depend not only on the size of the structure but also on layer thickness and layer exposure time. While this could still be competitive with standardised grinding and polishing processes, AM still has downsides when used for the manufacture of precision large-scale optical components, where several tens of minutes or even hours (depending on the requirements) are needed.

The need for faster printing technologies led to the development of other innovative printing methods, such as those based on projection micro stereolithography (PμSL), which combine the synergistic effects of grayscale exposure and meniscus coating methods [17]. These methods enable the manufacture of optical components with subvoxel-scale precision (sub 5 μm) and deep subwavelength surface roughness (sub 7 nm) while still achieving high printing speeds (i.e. 24.5 $mm^3 h^{-1}$ [17]). Despite this, the 3D printing of an aspheric lens with a height of 5 mm and a diameter of 3 mm took several hours to complete. Thus, despite the good resolution performance and speed enhancement, the time required to fabricate a single lens was still high. Furthermore, the lens recoating process also takes time to execute, further increasing the processing time.

One of the solutions proposed to accelerate the printing process was the use of micro continuous liquid interface production (μCLIP) technology [18]. For that, the authors of [18] used a custom-made polydimethylsiloxane (PDMS) thin film to maintain good surface roughness during the μCLIP printing of the optical components. This enabled the manufacture of millimetre-sized lenses with a surface roughness of 13.7 nm. Regarding the printing speed, the CLIP process reached a value of 4.9×10^3 $mm^3 h^{-1}$; this corresponds to a 200-fold improvement compared to the PμSL reported in [17].

An improvement in the speed of the printing processes while maintaining the surface quality of the component has also been demonstrated through volumetric 3D printing [19]. This technology can be divided into three categories: holographic exposure [20], orthogonal superposition [21], and tomographic volumetric printing (TVP) [22]. The latter has already been explored in the manufacture of optics, namely, to produce complex-shaped lenses [23]. To compare the performance of TVP with those of other printing methods, such as SLA and digital light processing (DLP), Peng *et al* [23] used SLA, DLP, and TVP technologies to print lenses. Later, the surface roughness of the lenses was evaluated using scanning electron microscope (SEM) images. The results of their work can be observed in figure 2.5.

As can be seen in figures 2.5(g) and (h), the TVP technologies enable the manufacture of lenses with good surface quality. The 3D lenses exhibited printing deviations from the designed lenses, ranging from 2.2% to 8.0%. Furthermore, no typical 'staircase' effect is visible in the SLA and DLP samples shown in figures 2.5 (e) and (f), respectively. The R_a of the TVP lenses achieved exceptional smoothness, ranking among the highest performances seen in AM competitors, with a value of 0.3340 nm. Furthermore, the results demonstrated the capability to print lenses at

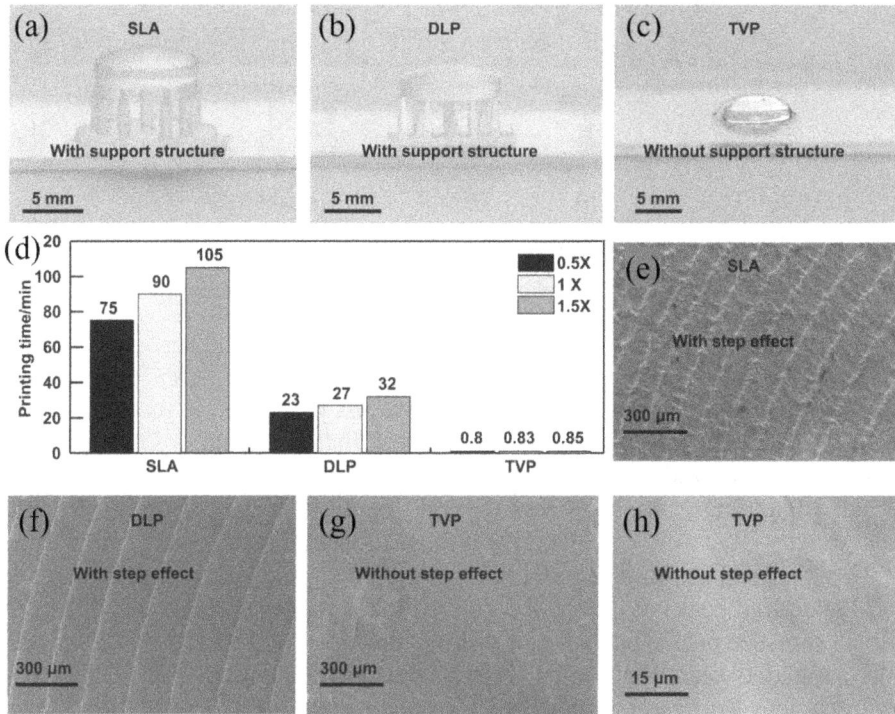

Figure 2.5. Representative CAD models of the same lens for different printers: (a) SLA, (b) DLP, and (c) TVP. (d) Fabrication times of the scaled lenses: 0.5× (16.1 mm^3), 1.0× (128.6 mm^3), and 1.5× (434.1 mm^3). SEM micrographs of lenses printed using (e) SLA, (f) DLP, and (g), (h) TVP before meniscus coating. Reproduced from [23]. ©2023 The Author(s). Published by IOP Publishing Ltd on behalf of the IMMT. CC BY 4.0.

high speed (see figure 2.5(d)), resulting in an ultrafast printing time compared to other methods. Specifically, a printing speed of approximately 3.1×10^4 mm^3 h^{-1} was reported, significantly surpassing its competitors [17, 18]. The technology also demonstrated the ability to print lenses with highly complex shapes, which is another key aspect of this research. Furthermore, another interesting characteristic of this method is that the lens requires the use of high-viscosity resins, allowing objects to remain in suspension during printing. This avoids the use of support structures that could otherwise compromise the lens, as demonstrated in [12].

TPP offers enough resolution to print lenses with high surface quality [24]. However, as stated earlier, this technology has limitations imposed by the duration of the printing process, which makes it impractical to manufacture lenses at the mesoscale. PµSL 3D printing [25] is another possible candidate, but it relies on extremely complex setups to guarantee surface quality and good resolution. Considering that the new masked stereolithography (MSLA) printers have updated 4 K resolution screens with pixel resolutions of around 35 µm, Nair *et al* reported the use of a fibre optic taper (FOT) on top of the liquid crystal display (LCD) screen to demagnify the images on the screen and print high-resolution lenses at the mesoscale (100 µm to 5 mm) [26]. To achieve this, the FOT was positioned between the LCD

Figure 2.6. SLA 3D-printed biconvex aspherical lenses. The lens diameter is approximately 50 mm. Reprinted with permission from [27]. © 2023 Optica Publishing Group.

screen and the resin vat, and the fabrication process proceeded in the same manner as in a standard MSLA process. To understand the advantages of this approach, lenses printed with and without the FOT were compared. From the results, it was concluded that the printing process through the FOT reproduced the lens shape almost exactly. Regarding the imaging tests at the lens's focal point, it was verified that when the FOT was not introduced into the system, the fabricated lens presented several undesired peaks at the focal point, while for the lens printed using the FOT, the focal spot-size profile had a single well-defined peak of 3.85 μm, which was close to the simulated peak (2.93 μm). Overall, the introduction of the FOT into the printing system led to improvements in surface roughness and resolution, demonstrating the potential of using a FOT to enhance the quality of the printing process at low cost while maintaining the speed of the MSLA processes.

The 3D printing of nonspherical surfaces is a key feature of 3D printing. In this regard, Aguirre-Aguirre and co-workers have documented the fabrication of both spherical and aspherical lenses [27] using the Form 3 SLA printer made by FormLabs® and Clear Resin from the same manufacturer as the host material. A photo of some of the lenses produced is shown in figure 2.6.

The results presented in [27] revealed lenses with smooth surfaces, showing errors of less than 2.5% in the radius of curvature, the optical power, and the focal length. For the aspherical lenses, it was verified that fabrication errors did not affect their performance. Regarding the biconvex spherical lenses, the authors verified a maximum resolution of 50.8 lpi, demonstrating their feasibility for general-purpose light-focusing applications such as solar concentrators and LED optics. However, despite the poor resolution for imaging applications, the authors still demonstrated the lenses' performance in an ophthalmologic application, obtaining blurry yet comparable images to the one observed with a standard lens.

2.1.2 Freeform lenses

The team of Aguirre-Aguirre has also explored the use of freeform lenses in another study related to the fabrication and characterisation of Alvarez lenses [28]. These lenses are designed explicitly for focusing applications and are complementary to each other, sharing a third-order polynomial surface. When aligned front to front, a light beam passing through them is unaffected. However, a lateral shift between the lenses allows for a change in the back focal length (BFL) of the system, resulting in a multifocal system, as illustrated in figure 2.7(a). These lenses offer a significant benefit in terms of correcting aberrations, reducing distortions such as coma, and minimising chromatic and astigmatic aberrations, especially in thermal lenses used in laser systems. Additionally, since the lens movement is achieved through a lateral shift of the lenses, this approach can be applied in compact systems while maintaining their varifocal ability [29]. An example of Alvarez lenses produced through SLA 3D printing technology, after undergoing post-processing methods such as polishing, is shown in figure 2.7(b).

The surface shape of the lenses presented in [28] was validated through the implementation of an off-axis null screen test. The results showed that the RMS discrepancies in the design parameters and the 3D printed element were 0.13 mm, and the peak-to-valley value was 0.7 mm for one set of the tested lenses. To justify the results, the authors claim that the post-processing of the lenses could be the reason for the poor-quality results. Despite this, their work effectively demonstrated the overall concept of an Alvarez lens and showed that high-precision 3D printing technologies, such as TVP, could be a viable option for reproducing this type of lens concept.

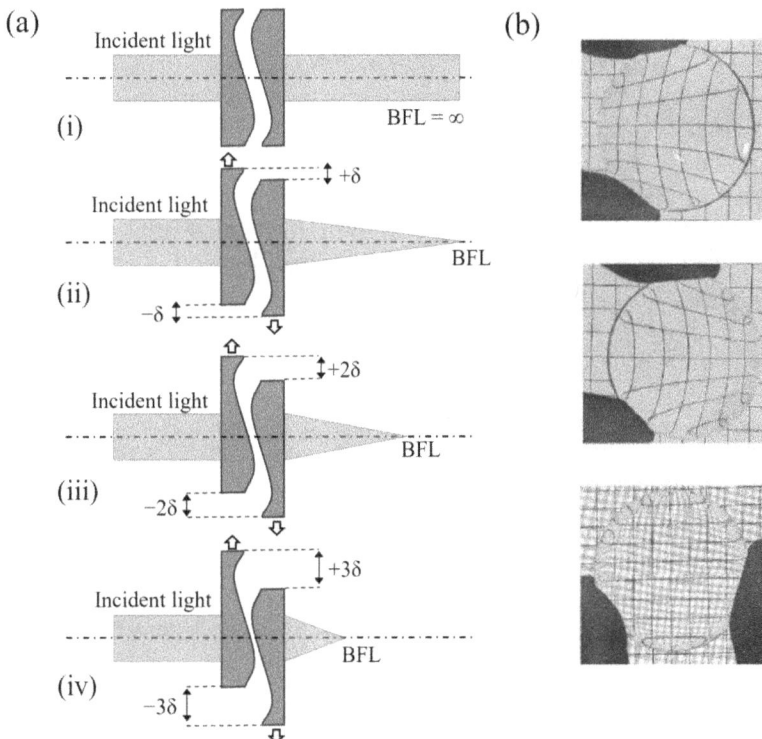

Figure 2.7. (a) Side-view representation of light passing through two Alvarez lenses. The lateral shifts of the lenses represented from (i) to (iv) allow control of the BFL. (b) Photos of the two halves of the Alvarez lens (top and middle images) and the assembled lens (bottom). Reprinted with permission from [28]. © 2024 Optica Publishing Group.

In the field of AM, the research focus has been expanding to the manufacture of optical components made of glass materials. The reason for this is the excellent properties of glass, including exceptional optical performance, high thermal resistance, and high chemical resistance [30]. Several 3D printing techniques have already been explored for the manufacture of optical components made of glass. Fused deposition modelling (FDM) [31] and TPP [32] have already been employed; however, these technologies present certain limitations. In the case of FDM, the printed components have low resolution, whereas in the case of TPP, the printing process is too slow to complete, thereby limiting the size of the components that can be printed. Therefore, new options have been explored for printing glass components, such as SLA and DLP, in conjunction with sol-gel mixtures [33] and glass slurries [34] as printing materials. The fabrication of glass aspheric lenses using a DLP 3D printer has already been reported in the research literature [35]. For this research, the authors used a glass slurry material composed of hydroxyl-ethyl methacrylate (HEMA), polyethylene glycol diacrylate (PEGDA), diethyl phthalate, and amorphous silica nanoparticles (Aerosil OX50). The lens model had a 14.6 mm radius and a convex part with a height of 1.1 mm. This was 3D printed with a layer thickness of 50 μm. Due to this layering process, a thin layer of uncured slurry was spin-coated onto the lens surface and subsequently photopolymerised, resulting in a reduction of the layering effect. Subsequently, the printed part underwent debinding (at 600 °C) and sintering (at 1250 °C) processes through heat treatments, transforming the 'green' 3D-printed lens into a pure glass aspheric lens. The results showed a lens with a smooth surface and an RMS value of less than 15 nm, exhibiting minimal deviations in the profile, with a maximum deviation of 170 μm. The imaging resolution was limited to 45.3 lp mm^{-1}, likely due to the high layer thickness used, indicating that there is still room for further improvement. Even so, this work paved the way for the fabrication of glass lenses.

Precision polishing, precision machining, and precision moulding are well-established technologies that are mandatory when enhancing the surface quality of standard lenses. Keeping this in mind, it is also easy to understand that to make 3D printed lenses ready for the market, it is necessary to guarantee high-precision surfaces. While this could be easily achieved for TPP, TVP, and MJ without the requirement for further post-processing technologies, they still face challenges related to the size and/or the elevated costs of their systems. Vat polymerisation techniques, on the other hand, are inexpensive and can facilitate the mass production of custom lenses. However, they are inherently affected by the layering phenomenon along the printing direction, as well as by pixel size. Thus, as discussed earlier, some studies have already employed post-processing techniques that can remove discontinuities along the printed parts. Examples capable of reducing the discontinuities associated with pixel size can be solved using grayscale photopolymerisation methods [17, 36, 37], allowing the pixel aliasing to be blurred and the curved contour to be smoothed in one layer. Demagnification of the projected image through FOT [17] is also a possible solution, despite the constraints related to reducing the build size. Meniscus coating [17, 38, 39] is a common post-processing technique used to reduce the layers along the printing direction, but this cannot eliminate the step changes on the printed object. Grinding, polishing, and glass curing are also possible solutions, but they are time-consuming and offer no advantages compared to the established methods for the lenses produced so far. On the other hand, the post-coating process [12, 40] has shown promising opportunities. In this regard, Shan *et al* [41] demonstrated the use of unfocused image projection and precision spin coating to obtain smooth surface 3D-printed lenses. For demonstration purposes, the authors used an MSLA printer and fabricated lenses of various sizes (3–70 mm) and profiles, including aspheric and axicon lenses. Some of these are displayed in figure 2.8.

As ultraviolet (UV) light passes through the LCD panel, it is obstructed by the boundaries of the LCD transistors. This manifests as a grid-type shadow, contributing to an uneven light intensity distribution, which ultimately produces staircase defects orthogonal to the printing direction. A study by Shan *et al* [41] demonstrated that manipulating the LCD screen's distance in the printing direction can lead to the projection of an unfocused image pattern, thereby reducing the pixelation effect and enhancing the smoothness of the printed lenses. Furthermore, the authors developed a precision spin coating process by integrating a mathematical model to predict and

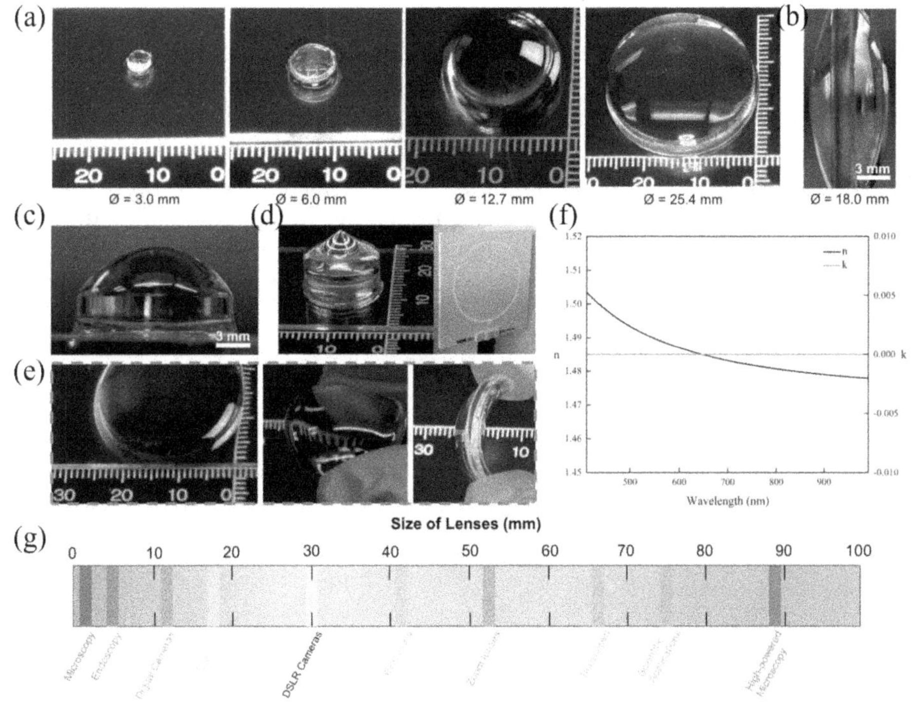

Figure 2.8. MSLA lenses printed by unfocused image projection and assisted by precision spin coating: (a) lenses of different sizes; (b) biconvex lenses; (c) aspherical lens; (d) axicon lens; (e) PDMS lenses. (f) Refractive index dispersion curve of the high-clarity resin material used for the 3D lens printing. (g) Applications of the different-sized lenses that the proposed method can produce [41] John Wiley & Sons. © 2024 The Author(s). Advanced Functional Materials published by Wiley-VCH GmbH.

control the process, enabling them to address lateral and vertical stair-stepping defects. After this, curing was performed under UV radiation in a vacuum chamber. The printed components produced by this technique exhibited excellent surface smoothness (<1 nm), excellent profile accuracy (<1 μm), and good reproducibility. Additionally, the printed convex lenses achieved a maximum modulation transfer function resolution of 347.7 lp mm^{-1}. These results demonstrate the promise of this technique for manufacturing complex optical lenses. The proposed method contributes to solving one of the main problems in AM, namely, the ability to print optical components at high speed while maintaining good surface quality.

2.1.3 Fresnel lenses

Fresnel lenses were first reported in 1822 by the French physicist Augustin Fresnel [42] and were first used in lighthouses as collimators. Fresnel lenses are made of concentric rings that approximate the curvature of a conventional lens. Depending on the optical application, they can focus or collimate a light beam in a manner similar to a regular lens. Overall, their design reduces the lens thickness and, consequently, its mass and volume, making this one of their main advantages. This produces a much thinner lens, particularly for large apertures. A schematic of the lens transformation is shown in figure 2.9.

Fresnel lenses have several applications in different fields, such as light collimators [43], photovoltaic panels [44], focusing elements in ultrasonic devices [45], acoustic lenses for medical imaging diagnosis devices [46], and automobile stoplights [47], among others [48].

Nowadays, there are several methods for producing Fresnel lenses, such as etching and layered deposition, injection moulding, compression moulding, and casting [49]. However, there are still limitations in terms of design features and

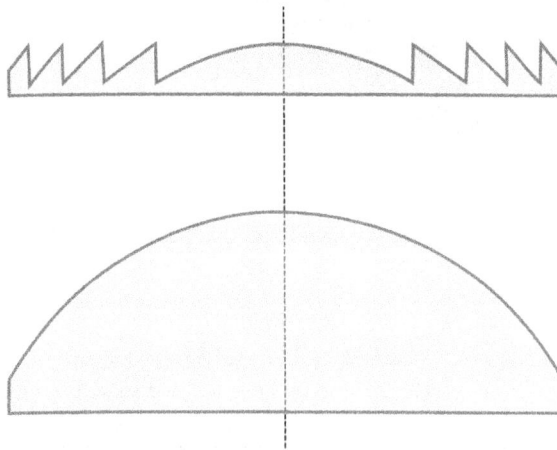

Figure 2.9. Schematic of a lens thickness reduction. At the bottom is a conventional lens, and at the top is a Fresnel lens.

process flexibility [50]. Regarding material usage, poly(methylmethacrylate) (PMMA) has been the most preferred choice for producing these lenses, as it allows for mass production via injection moulding due to its optical transparency and good mechanical properties.

One alternative that is beginning to be explored for manufacturing Fresnel lenses is AM. This method is a serious competitor to traditional techniques, as it enables the fabrication of lenses in a cost-effective manner, offers greater flexibility in the manufacturing process, and provides improved reliability. Additionally, it is possible to explore the use of different materials for the manufacture of these lenses.

The fabrication of a Fresnel lens begins with the design and calculation of its parameters, including the number of concentric rings, the focal length, and the bending angle (α_i) of each ring. Furthermore, the width of the rings (Δd) and the height of the ith ring (h_i) also need to be considered. To understand the meaning of each of these variables, refer to figure 2.10, which represents a Fresnel lens with all the parameters identified.

In figure 2.10, the bending angle (α_i) is given by [51]:

$$\tan(\alpha_i) = \frac{r_i}{n_r}\sqrt{r_i + f^2} - f \tag{2.1}$$

where r_i is the ith annular groove radius, n_r is the refractive index of the lens material, and f is the lens focal length. Here, Δd is given by:

$$\Delta d = \frac{r_l}{n} \tag{2.2}$$

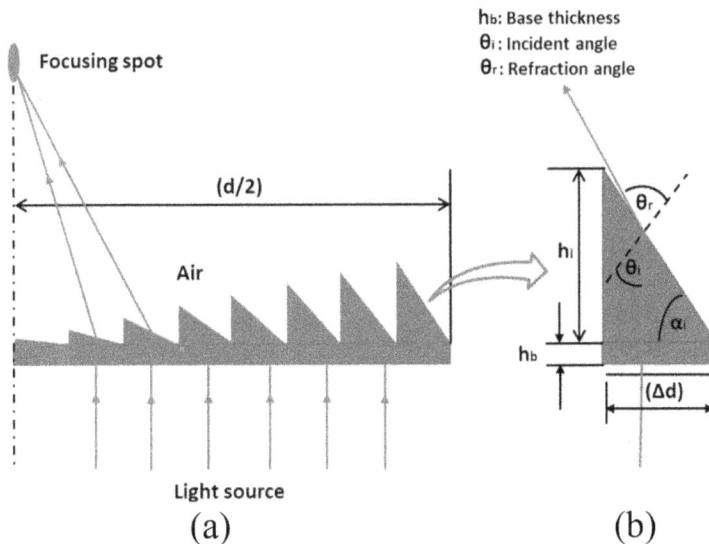

Figure 2.10. (a) Design of a Fresnel lens. (b) Detailed view of a Fresnel lens groove [51] John Wiley & Sons. © 2022 Wiley-VCH GmbH.

where r_l is the Fresnel lens radius and n is the number of rings. Using α_i and Δd, it is possible to determine the h_i value of the ith ring using:

$$h_i = \Delta d \times \tan (\alpha_i). \tag{2.3}$$

An example of an AM Fresnel lens with customised designs and properties was proposed in [52]. For this purpose, the authors utilised a DLP 3D printer and a commercial transparent resin, sold by ASIGA under the trade name DentaClear. They were designed with a 25 mm diameter and five ring zones, each with a constant height of 1.2 mm, and with a focal length projected to be 38 mm (approximately 1.5 times the diameter of the lens). The authors also included inks (in amounts that do not affect the lens transmission) for the purpose of selective colour filtering. The printed lenses, together with their transmission graphs for different light polarisation angles, are visualised in figure 2.11.

Without ink, the lenses presented optical transmittance values of over 90%. With the ink, the transmittance was slightly diminished, and certain attenuation bands appeared at specific wavelengths, depending on the colour of the ink. Moreover, the authors showed that collimated light passing through the lenses was effectively focused at a common focal point. These results demonstrate the capability of DLP technology to print tinted Fresnel lenses with dual functions, namely focusing and selective colour filtering.

Recently, developments in 3D printed Fresnel lenses have included the development of a 4D Fresnel lens [53]. This is of special interest because 4D structures pave the way for advancements in their capabilities, namely by changing their shape, properties, or functionalities in response to an external stimulus. The lens was fabricated using a DLP 3D printer, and three different thermochromic pigments with three distinct colours (blue, green, and red) were mixed with the resin prior to the printing process. After fabrication, the optical transmittances of the Fresnel lenses with and without the pigments were evaluated. The results of this characterisation are illustrated in figure 2.12.

As can be seen in figure 2.12(a), the addition of the thermochromic pigments significantly diminishes the optical transmittance of the transparent resin. However, in addition to the intrinsic material losses (resin and pigments), the thermochromic pigments exhibit temperature-dependent transparency, reaching maximum transparency at a specific temperature threshold [54]. Comparing the light transmittance of the printed lenses (figure 2.12(b)) with that of the polymer material, an overall transmittance reduction is observed. Yet, this was related to the light scattering along each of the 3D-printed layers.

Temperature tests were conducted with the 3D-printed lenses between 25 °C and 32 °C, demonstrating that transmittance increases with temperature, making them suitable for thermal sensing applications. Moreover, the results also showed that the focusing capabilities of the lenses were similar to those predicted theoretically. However, there was room for improvement in the resolution of the printed lenses.

Within the same topic and building on the achievement of the 4D Fresnel lenses described in the previous study [53], Ali and co-workers were able to report the

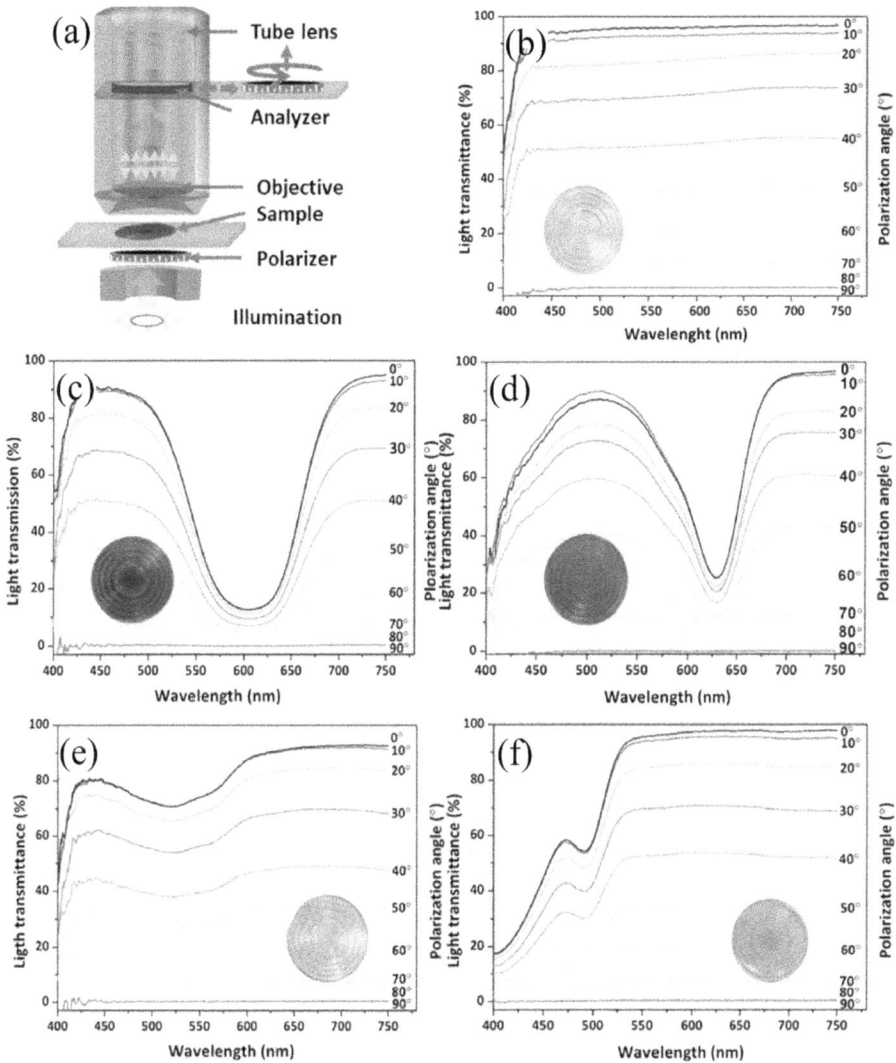

Figure 2.11. (a) Setup used to characterise the spectral transmission of the ink-coloured Fresnel lenses described in the text. Transmission spectra of the Fresnel lenses (b) without and (c–f) with colourants. Reprinted from [52], Copyright (2021), with permission from Elsevier.

fabrication of a lens with an additional dimension [55]. The fabrication process followed the procedure described earlier, utilising an MSLA printer to fabricate the lenses. The resin consisted of a mixture of HEMA and PEGDA in a 1:1 ratio, along with trimethylbenzoyl diphenylphosphine oxide (TPO) at 2.5%. These acted as the monomer, cross-linker, and photoinitiator, respectively. As in [53], thermochromic pigments were added to the resin mixture to make the resin temperature sensitive. To add another feature to the lens, a holographic diffraction pattern was incorporated into the flat surface of the Fresnel lens. This was done through the attachment of a

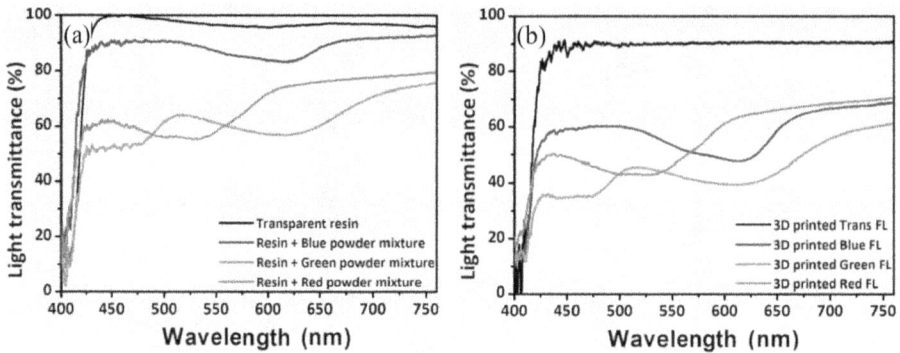

Figure 2.12. Optical transmittances of the polymer material: (a) before the printing and (b) after the printing of the Fresnel lens. Reprinted from [53], Copyright (2022), with permission from Elsevier.

microsized holographic relief polyvinyl chloride (PVC) film pattern to the 3D printing building plate, giving the first printing layer (flat lens side) the same holographic pattern. By doing this, it was possible to equip the lens with mechanoluminescence sensing capabilities, thanks to its holographic patterning, which made it only necessary to add an image sensor. Regarding the focusing properties, the lenses were characterised in the visible region, exhibiting a focal length with an average deviation of less than 2 mm.

More recently, Ali *et al* were also able to add another feature to Fresnel lenses, showing the possibility of using them for the measurement of varying concentrations of ethanol, isopropanol, and methanol in alcohol sensing [56]. The mechanism behind the measurement scheme was associated with the polymer's swelling effect [57, 58]. This affects the transmission intensity [57, 59], allowing the authors to report sensitivities to ethanol, isopropanol, and methanol of 0.37, 0.33, and 0.24 μW/vol.%, respectively. In addition, the sensors presented a detection limit of 5 vol.% and a response time of 25–30 min, showing competitive results compared to those described in the research literature [60].

2.1.4 Contact lenses

Vision problems affect millions of people worldwide, and contact lenses are medical devices widely used to correct these problems. Contact lenses are directly placed on the cornea of the eye as an alternative to glasses. Due to their use in such sensitive areas, they must be made from materials that fulfil strict requirements. The criteria to be met include general aspects related to patient comfort, along with optical, biophysical, and chemical standards [61, 62]. Other important properties include gas permeability, biocompatibility, optical transmittance, surface wettability, and mechanical properties. Materials that meet these criteria and are being used in the manufacture of contact lenses include poly(2-hydroxyethyl methacrylate) (pHEMA) [63], glyceryl methacrylate [64], polyvinyl alcohol [65], PMMA [66], fluoro-siloxane acrylate [67], cellulose acetate butyrate [68], and perfluoroethers [69], among others.

The choice of material, however, is dependent on the type of contact lens required (e.g. rigid, soft, or hybrid) for a specific application.

Contact lenses are usually hemispherical and are commonly manufactured by spin casting and cast moulding. These processes are expensive, since they need specialised machines to produce the moulds. Furthermore, these methods have a low degree of freedom, as a change in one lens parameter requires the manufacture of individual expensive moulds [70–72]. The lathe-cutting method [73] is another approach used in the manufacture of contact lenses. It has normally been used to manufacture specialised lenses due to its high precision and customisation. However, it can be considered time-consuming and costly, being only suited for small to medium-scale productions. Furthermore, since each lens is sculpted separately, reproducibility issues may arise.

Within contact lens research, one technology that is gaining interest is smart lenses. The idea behind these lenses is to take advantage of the interaction between the contact lenses and the eye tissues and tears, allowing the monitoring of diseases such as dry eye disease and intraocular pressure, among others [74–76], enabling patient comfort and timely quantitative analysis for point-of-care diagnostics. Usually, smart lens manufacture is achieved through the use of nanostructures. This can be complex and challenging to execute in commercial contact lenses [77].

AM allows for increased design freedom in lens production, enabling the creation of lenses with various shapes and dimensions at a relatively low cost. Considering these opportunities, contact lenses have also drawn the attention of the AM community to this topic. One related study is described in [78], which reports the production of contact lenses and smart contact lenses via DLP technology. Today's lenses can have several colours, allowing them to serve aesthetic purposes and sometimes to correct colour blindness. As a result, in [78], different nontoxic food-grade colours were mixed into a liquid monomer resin before the printing process. The food pigment was incorporated into the photopolymerisable resin at a concentration of 2% by total volume without compromising its transmittance. The lenses were designed and successfully fabricated with intricate microchannels in their borders. These were intended to act as optical transducers [79] in biosensing applications, specifically through the observation of changes in microchannel geometry using images captured by external cameras.

Following the printing process, it was verified that the lenses suffered from the common AM layer effect. Furthermore, the removal of the lens from the building plate caused surface damage due to its strong adhesion to the plate, resulting in poor lens transmittance. Additionally, the rough surface of the building plate resulted in a poorly printed surface, which in turn led to highly scattered light transmission. To mitigate the problems associated with the layering effect, a coating was applied to the lens. Regarding the effects related to surface damage caused by removal and the rough surface plate, a smooth surface PVC film was added to the building platform prior to the printing process. The optimisation process enabled the authors to achieve an R_a of ≈ 10 nm. The results, namely the SEM images and the roughness of the contact lenses obtained through the optimisation process, can be seen in figure 2.13.

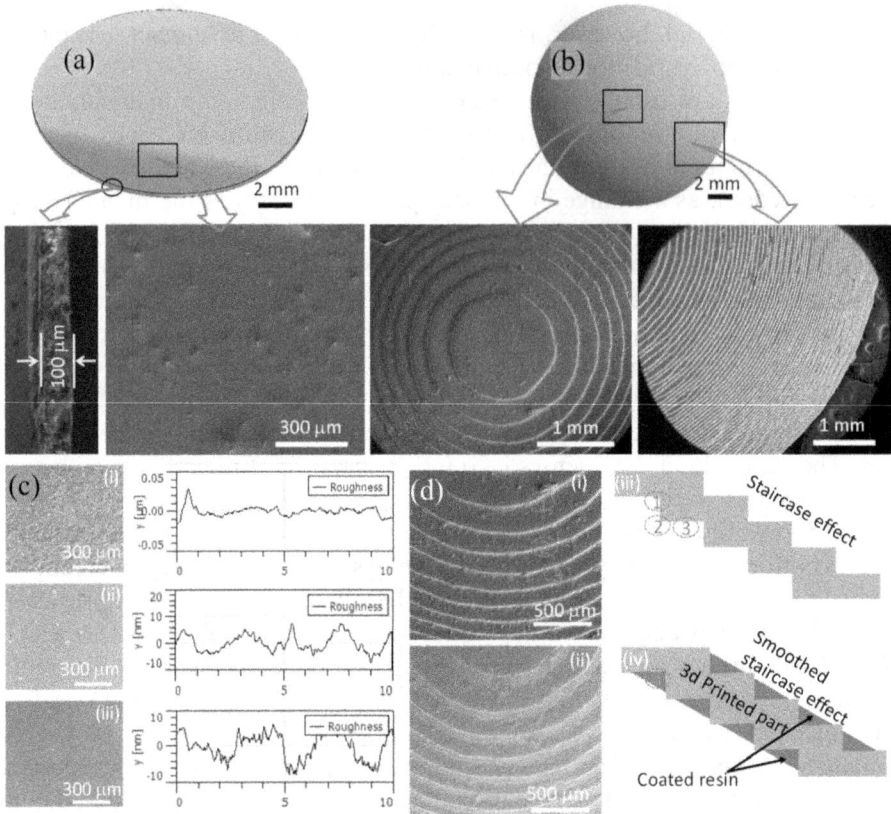

Figure 2.13. (a) SEM images of the top and cross-section of a flat disc. (b) 3D printed contact lens with respective top and side wall SEM images. (c) SEM images showing different lens manufacturing improvements: (i) direct from the building plate, (ii) with a resin coating, and (iii) after printing on a flat PVC film. The graphs on the right correspond to atomic force microscopy measurements. (d) SEM images (i) before and (ii) after the lens coating. The corresponding schematics on the right-hand side illustrate the surface quality enhancement resulting from the layering effect. Reproduced from [78]. CC BY 4.0.

Returning to [78], the authors of this article implemented a nanopatterning process at the lens surface through holographic laser ablation, specifically by using direct laser interference patterning (DLIP). For that purpose, a synthetic black dye was added to the surface of the lens. This enhanced the interaction of the laser with the material of the lens, allowing the surface to be patterned through ablative processes, namely the constructive interference phenomenon [80]. Through this methodology, a holographic pattern was imprinted on the lens surface, enabling it to function as a transducer for sensing ocular parameters. A representative schematic of the implemented fabrication process and the resultant SEM image of the holographic pattern are shown in figure 2.14.

Colour vision deficiency (CVD), or colour blindness, is a condition that affects humans, preventing those affected from distinguishing certain colours [81]. So far, there is no cure for this condition; thus, available approaches focus on enhancing the

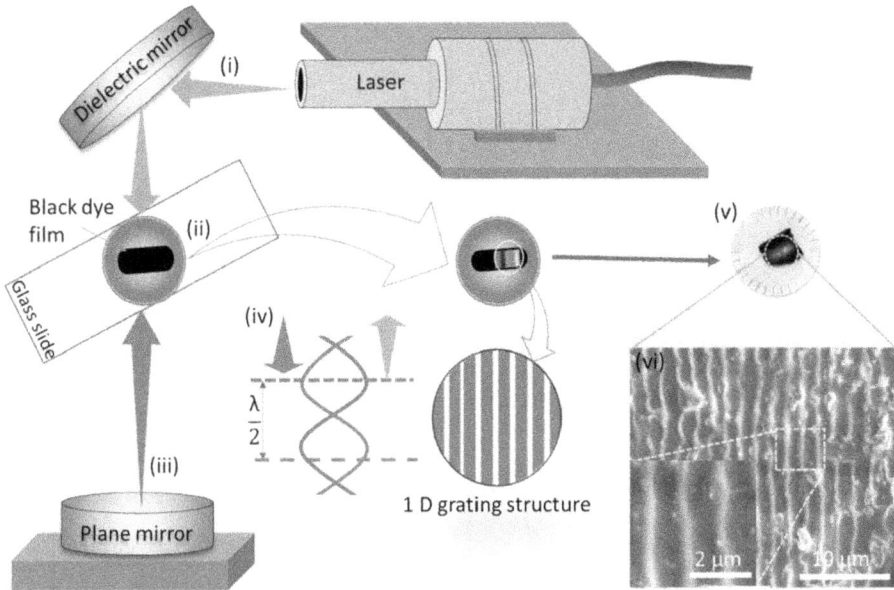

Figure 2.14. Representative schematic showing the fabrication of the nanopattern using DLIP in reflection mode and its different steps: (i) the laser beam is guided to the lens surface, (ii) passes through the lens with the black dye, and (iii) reflects at a mirror placed below the lens. Interference ablates the lens surface, creating (iv) a 1D grating structure. (v) Digital photograph of the hologram projected in the lens, and (vi) SEM image of the holographic grating. Reproduced from [78]. CC BY 4.0.

patient's colour perception through the use of CVD glasses or contact lenses [82]. The idea of tinting the glass with shades of colour was first introduced in 1837 by Seebeck [83], where CVD patients were able to distinguish between the relative brightness of red and green shades using a red filter followed by a green filter, enabling independent and individual activation of photoreceptor cone cells, and with that, proper recognition of the signals by the brain and better colour perception. Since then, most of the proposed solutions have followed the same principle, and different materials have already been tested [84, 85].

Taking into account the opportunities offered by AM in the development of contact lenses, Haider *et al* [86, 87] have already proposed the 3D printing of contact lenses for the treatment of colour blindness. For this purpose, he used a synthesised pHEMA resin tinted with wavelength-selective fluorescent filter dyes, Atto 565 and Atto 488, which are considered harmless to corneal and epithelial cells [88]. These dyes have absorption bands corresponding to wavelengths in the ranges of 550–580 and 480–500 nm. The lenses were manufactured using SLA technology, and the results showed that surface smoothness depended on the printing angle, as described in [12], specifically in the vertical orientation (perpendicular to the lenses' principal axes). The best results were achieved with a vertical orientation perpendicular to the lenses' principal axes. Optical test results for commercially available glasses used in the treatment of CVD and the 3D-printed tinted contact lenses are presented in figure 2.15.

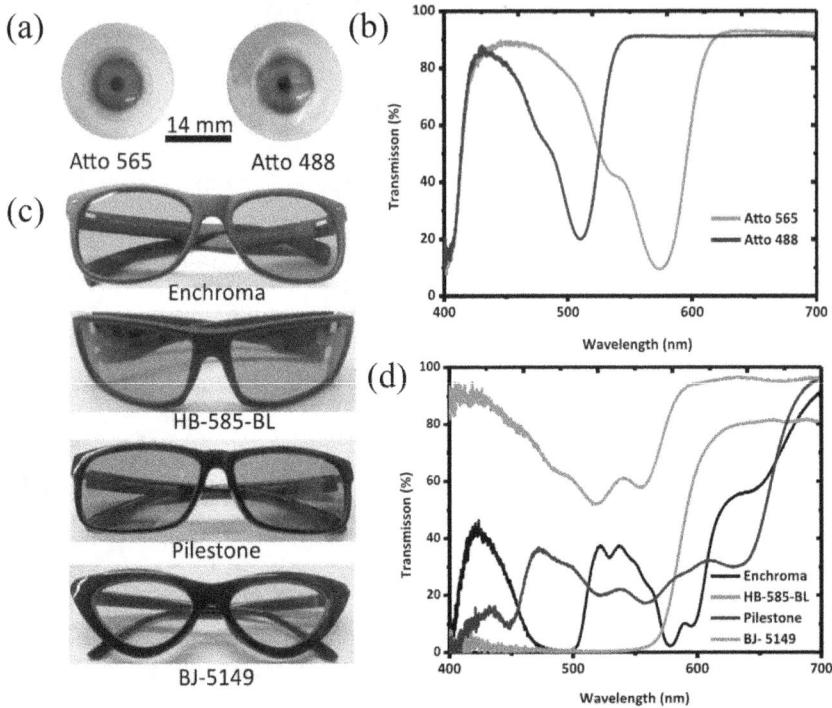

Figure 2.15. (a) 3D-printed tinted contact lenses placed on an eye model. (b) Transmission spectra of the tinted contact lenses. (c) Commercially available glasses for colour blindness. (d) Transmission spectra of the commercially available colour blindness glasses. Reprinted from [86], Copyright (2022), with permission from Elsevier.

The 3D-printed tinted contact lenses shown in figure 2.15 demonstrated generally improved transparency compared to the commercial eyeglasses, as well as enhanced unwanted wavelength-blocking capabilities, reaching values of 80%–90% [86]. The utilisation of these lenses by patients revealed improvements of up to 43% in identifying Ishihara testing plates, which conclusively confirmed the high efficacy of the lenses and the potential of using AM to create technologies that improve people's quality of life.

While contact lenses are considered a significant advancement in addressing ophthalmologic issues in a minimally invasive manner, their regular use can lead to various conditions, such as eye dryness, eye inflammation, or general eye discomfort [89]. Various solutions have been proposed to address these issues, including the use of eye drops, electrical stimulation of the lacrimal gland, and the insertion of a punctum [90, 91]. While these techniques help hydrate the eyes, they are not effective in maintaining moisture for extended periods. To prevent moisture evaporation, more advanced technologies, such as graphene coating [92] or the use of electro-osmotic flow [93], have been tested. Even so, these technologies are too complex and time-consuming, making them impractical for large-scale production. One promising solution is found in AM, which enables the fabrication of microchannels within the lens material. This has already been tested through the fabrication of a mould containing microcapillary channels for moisture retention [94]. The concave and convex mould parts were fabricated using either FDM or SLA printers. The lenses were then fabricated via mould casting, where a soft PDMS elastomer was used as the reference material for the lenses. The staircase effect present on the moulds was transferred to the lens and acted as parallel microchannels, demonstrating its possibility for capillary rise. The printing orientation played a crucial role in the efficiency of the lenses, with printing angles orthogonal to the optical axis yielding the best results. Polishing of the lenses' central regions was also performed to enhance transparency in that area. Moreover, the lenses were exposed to oxygen plasma to make them hydrophilic. The resulting contact lenses promote self-moisturization, offering a simple and effective solution to the problem of eye dryness.

2.2 Mirrors

An optical mirror (see figure 2.16) consists of a reflective surface designed to direct, focus, or manipulate light in optical systems, ensuring precise imaging and beam control.

Mirrors are optical elements that play a crucial role in various optical systems, including telescopes, cameras, and scientific instruments. To ensure optimal performance, they must provide high precision, stability, and reliability— qualities that frequently require complex and expensive manufacturing technologies.

Nowadays, ultra-precision machining technologies, such as single-point diamond turning, direct polishing, and chemical and mechanical polishing [95–97], offer

Figure 2.16. Examples of optical mirrors.

exceptional accuracy. However, they are often costly and complex. Conversely, AM manufacturing provides a compelling alternative, as it allows the fabrication of complex designs with reduced material waste, lower costs, and faster production processes. These advantages are attractive for producing mirrors [98, 99].

In aerospace applications, telescopes are designed to achieve high levels of light collection; thus, mirrors, which form the core part of telescopes, need to have high apertures. However, this leads to increased weights, raising the launch cost. To solve this, mirrors can be designed with lightweight materials and different infill structures. The design of these structures can be complex, time-consuming, and costly. Thus, AM has been identified as an alternative to mirror fabrication.

2.2.1 Polymer mirrors

Mirrors can be fabricated from various AM materials, including metals, ceramics, and polymers. Polymers offer low weights, can be fabricated with low-cost 3D printing technologies, and are easily post-processed. An example of this printing capability was reported in [40], where an aspherical mirror intended to act as a focusing element in a dual-axis confocal microscope was produced through the SLA method, followed by thin metal evaporation on the mirror surface. The authors designed the mirror to have a focal length of 6 mm and an outer radius of 10 mm, with a 1.1 mm centre hole. The mirror was 3D printed as a solid piece and later smoothed with a thin coat of UV-curable liquid resin that filled the pores and depressions resulting from the layering effect, yielding a high-quality mirror with 70 nm RMS surface roughness. Finally, the surface was metallized to make it reflective. This was achieved through the deposition of a 150 Å titanium seed layer, followed by a 1000 Å aluminium layer. The resulting mirrors behaved as expected for flat aluminium across the 200–1800 nm wavelength range. Images of the fabricated mirror produced using various fabrication processes are shown in figure 2.17.

Despite being lightweight, which is beneficial, for instance, in space applications, the surfaces of 3D-printed mirrors require metal evaporation for their proper operation, meaning that the polymer must maintain its integrity under vacuum pressure. Additionally, polymers are well known for their poor mechanical strength, large thermal expansion coefficients, and low operating temperatures. While mirrors fabricated from polymers could still find niche applications, their use in high-resolution imaging systems, such as those found in space applications, cannot be

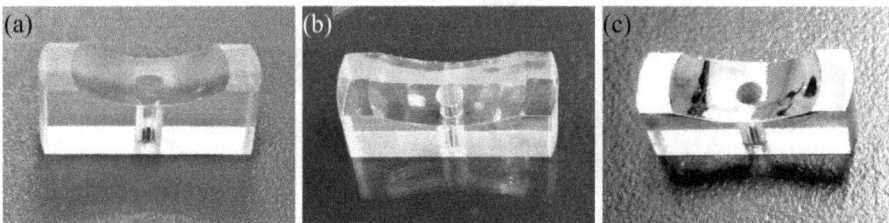

Figure 2.17. SLA 3D printed parabolic mirror at different fabrication stages: (a) as printed, (b) after smoothing, (c) after aluminium evaporation. Reproduced from [40]. CC BY 4.0.

considered. For these applications, the high mechanical and thermal performance of metal and ceramic materials is far more appealing.

2.2.2 Metal mirrors

Metals are one of the most widely used materials for producing mirrors. The reasons are associated with their high reflectivity, durability, and strength. Beryllium and aluminium are two materials commonly employed in the fabrication of metal mirrors. Examples include the 6.6 m aperture primary mirror of the James Webb Space Telescope [100], which is composed of 18 beryllium sub-mirrors, and the Wide-field Infrared Survey Explorer [101], which employed an all-aluminium design scheme, where aluminium mirrors were mounted in an aluminium structure. Despite their excellent performance, intense research has been directed towards the development of lightweight mirrors [102, 103]. These are mirrors where a percentage of the mass has been removed without adversely affecting structural integrity. While one could consider making a thinner mirror to reduce its weight, problems arise during the mounting phase, as distortions appear due to gravitational sag. Thus, thicker mirrors are preferable to maintain rigidity, while excess material can still be removed in locations where it is not needed. In the research literature, three categories of lightweight mirrors have been reported: contoured-back solid mirrors, open-back mirrors, and sandwich mirrors, as shown in figure 2.18.

Contoured-back mirrors are simple to fabricate but do not offer substantial weight reduction. Regarding open-back mirrors, hexagonal, square, triangular, and circular patterns have already been tested; however, they are challenging to fabricate using traditional techniques. Additionally, despite achieving a higher weight reduction, they exhibit less rigidity. On the other hand, sandwich mirrors are more rigid and offer high levels of weight reduction, but they are much more complex and expensive to manufacture [104].

Despite the opportunities offered by the conventional mirrors displayed in figure 2.18, the continuous requirements for lightweight designs with high resolution have driven research to further optimise the mirror design, namely through topology optimisation (TO). This relates to layout optimisation, which aims to find the optimal load path for a given boundary condition, thereby enabling the determination of the best material distribution within the part that fulfils certain constraints, such as compliance and displacement. The use of AM for the production of metallic

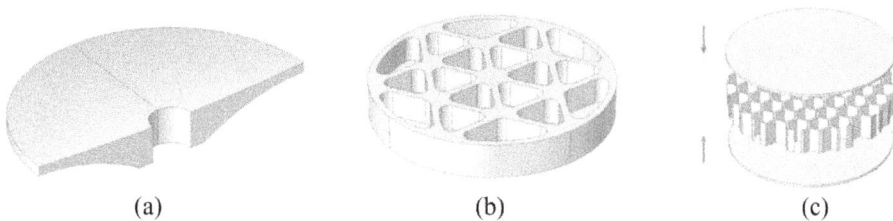

(a)   (b)   (c)

Figure 2.18. Schematics of three different lightweight mirror styles: contoured back (a), open back (b), and sandwich (c). Reproduced from [104]. CC BY 4.0.

mirrors adds flexibility in terms of design freedom; thus, complex structures obtained using TO can be easily processed, allowing creativity to play a part in mirror development [105]. Thus, AM virtually eliminates the need for a machine shop to produce an optical-quality mirror.

Selective laser melting (SLM), direct metal laser sintering (DMLS), and electron beam melting (EBM) are three common AM techniques used to produce metal mirrors. High-performance AM mirrors have already been reported in the literature. An example of these can be found in the study developed by Hilpert *et al* dedicated to the design and optimisation of lattices of Voronoi structures. This research resulted in mirror prototypes produced using SLM in aluminium alloy [106, 107], as depicted in figure 2.19.

Herzog *et al* [108] were among the first to report the additive manufacture of lightweight mirrors using TO methods. In this study, they reported the manufacture of two types of flat surface mirrors: one made of aluminium alloy (AlSi10Mg) and the other made of titanium alloy (Ti6Al4V). The mirrors were optimised to have high stiffness, be lightweight, and exhibit a minimum natural frequency of 250 Hz. For each material, two different designs were tested: one based on a top mount (figure 2.20(a)) and the other based on a side mount (figure 2.20(b)). The mirrors made of aluminium alloy were manufactured through DMLS, while those made of titanium were fabricated through EBM. After printing, the mirrors underwent a post-processing treatment that included a hot isostatic press and heat treatment to enhance their density, eliminate porosity, and reduce internal stresses. Later, the mirrors were also milled to reduce the time required for the subsequent grinding and polishing processes. The results obtained after the post-processing treatments are shown in figures 2.20(c) and (d) for the aluminium and titanium alloys, respectively.

The team then characterised the developed mirrors, showing that the post-processing treatments were effective for the aluminium alloy mirror but were unable to remove the porosity of the titanium alloy mirror. For the latter, the authors were

Figure 2.19. Prototype of a diamond-turned, additively manufactured aluminium alloy mirror coated in nickel phosphorous. The lightweight structure is based on Voronoi cells, where the density of the Voronoi cells depends on the mirror mount locations. Reprinted from [106], Copyright (2018), with permission from Elsevier.

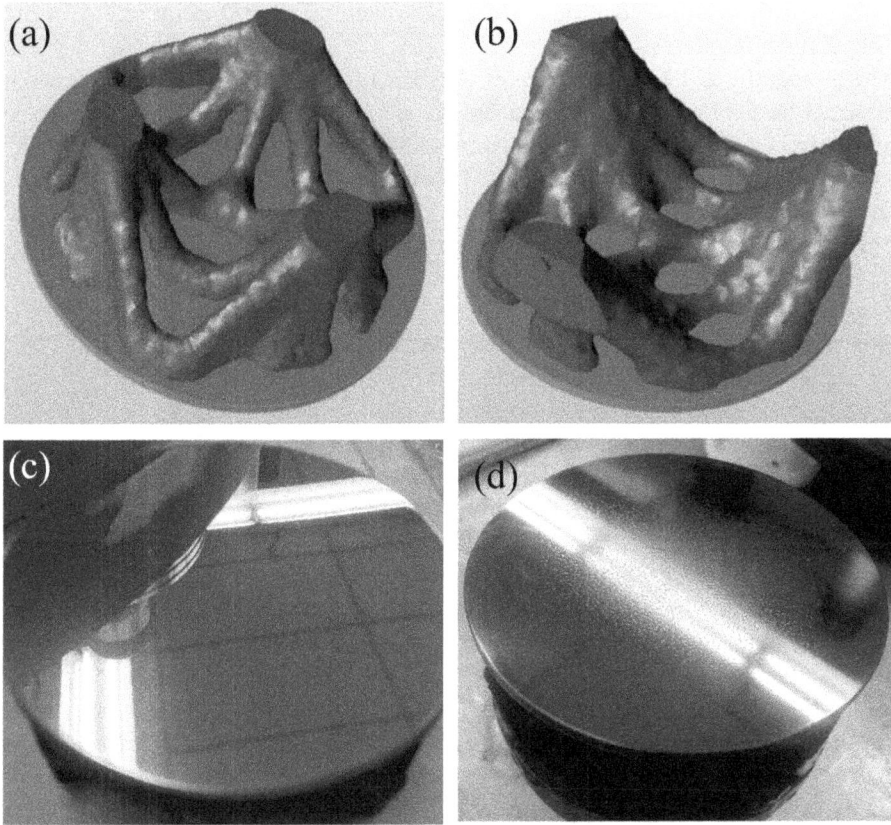

Figure 2.20. Mirror designs topologically optimised for high specific stiffness and low surface displacement: (a) top and (b) side models optimised for both AlSi10Mg and Ti6Al4V. The final results obtained for (c) AlSi10Mg and (d) Ti6Al4V are presented. The former does not exhibit defects, while the latter shows porosity defects visible at the centre of the mirror. Reproduced with permission from [108]. © (2015) COPYRIGHT Society of Photo-Optical Instrumentation Engineers (SPIE).

able to report a 22 nm RMS microroughness. While the mirror structure required improvements for use in space applications, the prototype showed the potential of AM in the development of TO lightweight mirrors.

Another application of TO mirrors fabricated through AM is found in reflective optics for high-energy laser systems, such as those used in materials and energy research, as well as in military applications, where systems delivering tens of kilowatts of power are typically employed. For these applications, the optical elements should have a low thermal expansion coefficient, high specific stiffness, and high thermal conductivity to dissipate heat effectively from the optical surface. Traditional systems rely on passive cooling and highly reflective coatings to withstand laser damage. With this in mind, Mici *et al* [109] explored different lightweight mirror architectures to allow active cooling of the mirror, namely through finite element analysis. These mirrors were optimised for both stiffness and thermal dissipation, demonstrating the potential of 3D printing to improve optomechanical systems.

More recently, a 100 mm diameter AlSi10Mg alloy mirror with a honeycomb structure was explored in the literature to examine its stability in air over long periods and in environments with constant changes in temperature [110]. The mirror was fabricated using the SLM process. To improve the as-printed mirror accuracy, the mirror underwent a resin replication process, in which a liquid epoxy resin with a curing agent was applied between the mirror and a glass mould of high optical accuracy. The results revealed a surface accuracy of 0.033λ RMS ($\lambda = 632.8$ nm) and good surface smoothness ($Ra = 1.3$ nm), which were attributed to the replication process. The mirror also showed excellent stability for 42 days in the open air as well as good dimensional stability for temperatures ranging between, with little stress, and exhibited good dimensional stability at 25 °C–35 °C.

While TO offers interesting opportunities for developing efficient and lightweight designs, it is necessary to consider the constraints that AM can impose when dealing with the 3D printing of intricate structures. During the printing process, each part of the object being created must be fully supported from below to ensure the quality of the final product. For complex TO structures, the use of support structures is normally required to prevent the structural material from being distorted or even collapsing and failing. The disadvantages of this approach are that the printing time increases, the material usage increases, and one also needs to consider the time required to remove the supports, the inherent difficulty in removing those supports from intricate structures, and the damage this can cause to the surface of the parts during support removal. Thus, it is beneficial to design mirrors that do not need support structures during their production; instead, self-supporting structures should be used. However, the inclination of downward-facing and non-supported (overhang) surfaces cannot be at too large an angle with respect to the print direction, as this can lead to deformed printed parts. Recently, Guo et al [111] reported a study of several infill structures designed to eliminate the need for support structures. To accomplish this, the authors developed a solution based on two different methods: moving morphable components (MMCs) [112] and moving morphable voids (MMVs) [113]. Numerical results revealed that the two proposed approaches had the potential to create mirror designs without overhang constraints.

Recently, a double-faced mirror with a TO-based self-supporting lattice was proposed, and the same design was applied as the infill structure of the mirror [114]. The two-sided mirror was manufactured through SLM using an aluminium–silicon alloy. Before the mirror was printed, several tests were conducted to verify the printer's limitations when printing the lattice structure. In these tests, the strut angle (α), related to the building platform, and the flat surface (bridge distance—Db) parallel to the building platform and supported by two arms (see figure 2.21(a)), were tested to verify the print limit values. The tests revealed that the SLM printer was capable of printing structures with bridge distances of up to 2 mm (see figure 2.21(b)). Regarding the strut printing angle, it was observed that the printer could produce high-quality structures with a strut angle of 40° or greater. For lower angles, the printed structure presented defects and lower-quality prints (see figure 2.21(c)).

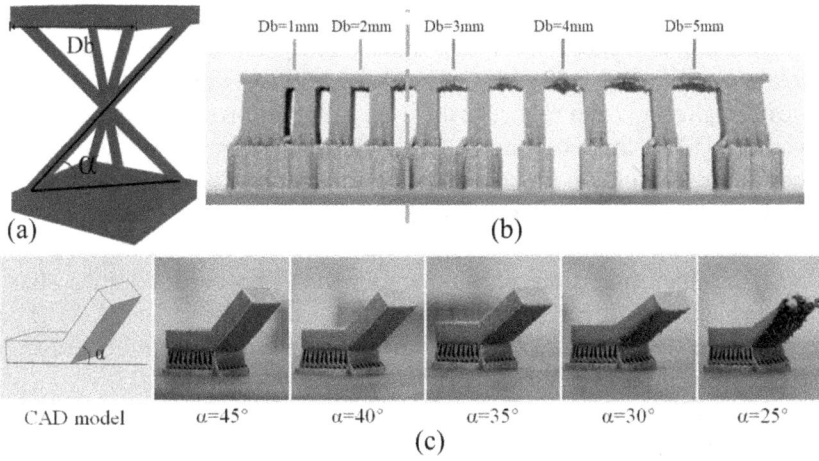

Figure 2.21. (a) Schematic of the self-supporting lattice structure, showing the critical parameters: vertical angle, α, and bridge distance, Db. SLM printing results for different: (b) bridge distances, and (c) vertical angles. Reprinted from [114], Copyright (2022), with permission from Elsevier.

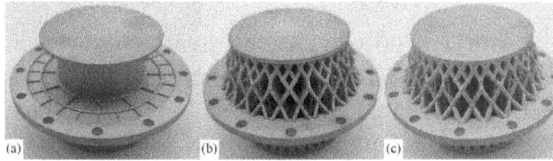

Figure 2.22. Double-sided mirror fabricated through SLM: (a) original mirror, (b) SG lattice infilled mirror, (c) SDG lattice infilled mirror. Reprinted from [114], Copyright (2022), with permission from Elsevier.

Regarding the strut diameter, the tests revealed that the strut structures can have a diameter of 0.8 mm. After parameter optimisation, the authors were able to print the double-sided mirror. For this, two model lattice structures were considered: the size–gradient (SG) self-supporting lattice structure design and the size and diameter–gradient (SDG) lattice structure design based on TO. The results obtained for each of the model lattice structures can be seen in figure 2.22 [114].

The results for the mirrors demonstrated that the mirror with the SDG lattice structure based on TO presented better stiffness, a higher first natural frequency, and high compression strength, and eliminated the need for support structures at a reduced material cost.

2.2.3 Ceramic mirrors

Although most reports have focused on the fabrication of metal mirrors, primarily due to their ease of manufacture and their potential for employing ultra-precision machining technologies such as single-point diamond turning, there have also been studies on the development of ceramic mirrors. For this, materials such as SiO_2 [115], Al_2O_3 [116, 117], ZrO_2 [118, 119], and bioceramics [120, 121] have been fabricated using SLA processes. Recently, silicon carbide (SiC), known for its excellent properties such as low density, high stiffness, high optical quality, low thermal expansion, and excellent dimensional stability, has attracted the attention of the scientific community [122, 123]. The fabrication of SiC parts with complex geometries has already been reported through colloidal processing techniques, such as slip-casting [124], tape casting [125], and gel casting [126]. However, these techniques share a lack of precision and are expensive and time-consuming to fabricate [127]. Furthermore, the fabrication of lightweight SiC mirrors, which is relevant for space-based applications, is challenging due to the complexity of their shape and the high-precision requirements. One proposed solution has been described through the direct 3D printing of SiC ceramic mirrors [128]. This was achieved through a combination of 3D printing, polymer burnout, pre-sintering, precursor infiltration, and pyrolysis steps, as depicted in figure 2.23.

The authors of [128] first prepared a low-viscosity and highly dispersed SiC slurry. This consisted of SiC nanoparticles, resin monomers, sintering additives, photo-initiators, and dispersants, creating a photosensitive SiC slurry ready for 3D printing. The fabrication was performed using an SLA 3D printer with a layer thickness of 50 μm, and sufficient time was allocated for each layer to polymerise the slurry. To produce the mirror, an optimisation procedure was first performed with a test sample.

After printing the samples, post-processing treatments were required. The first post-processing treatment consisted of polymer burnout, which was performed in an N2 atmosphere at 800 °C at a rate of 0.5 °C min^{-1}. Then, the parts were soaked for 2 h and subsequently cooled to room temperature. This resulted in a pyrolysed SiC body. Next, a pre-sintering process was conducted at 1800 °C (at a rate of 1 °C min^{-1}), followed by a one-hour soak, and then the sample was cooled to room temperature. This allowed the authors to produce a pre-sintered body. Finally, the last post-processing treatment consisted of precursor infiltration and pyrolysis. This was done to

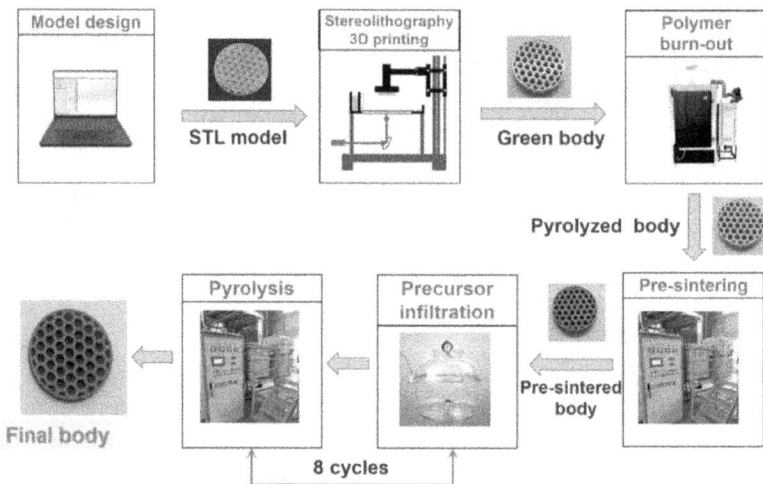

Figure 2.23. Flow diagram of the fabrication of a lightweight SiC ceramic optical mirror. Reprinted from [128], Copyright (2020), with permission from Elsevier.

improve the density and strength of the body. For this, polycarbosilane (PCS) and divinylbenzene (used as a solvent) were mixed in a weight ratio of 2:1. The pre-sintered body was then infiltrated with the solution under vacuum and cured at 60 °C for 1 h. The body was finally pyrolysed in an N_2 atmosphere at 1200 °C for 1 h. This process was repeated eight times to produce the final body. The corresponding SEM images of the cross-section and surface of the prepared SiC body during the different post-processing steps reported in [128] can be seen in figure 2.24.

As reported in [128] and shown in figures 2.23(a) and (b), the green body exhibits distinct layers resulting from the layer-by-layer AM process. Additionally, the printed parts exhibit a dense structure without pores or cracks, demonstrating the method's feasibility. After pyrolysis, the layering effect disappears, and the polymer and dispersants present in the structure are burned off, leaving a network of pores with SiC ceramic particles separated from each other (see figures 2.23(c) and (d)). With the pre-sintering process, grain boundaries are formed, and the ceramic becomes denser. Yet, it still shows porosity, indicating that the body does not reach full densification. After the eight-cycle pyrolysis process, the particles resulting from the pyrolysis of PCS fill the small voids present in the body. The surface appears denser, more solid, and void-free, as shown in figures 2.23(g) and (h), indicating the success of the technique and allowing the team to proceed with the production of the mirrors.

The SiC ceramic optical mirror was designed with a lightweight support structure, as seen in the standard tessellation language (STL) model presented in figure 2.25(b). Photos of the mirror taken during the various production steps (3D printing, burnout, pre-sintering, precursor infiltration, and pyrolysis) are shown in figure 2.25(c)–(j).

The SiC mirror results presented in [128] and summarised in the photographs shown in figure 2.25 revealed that 3D printing has sufficient accuracy to print the complex geometries necessary for the lightweight mirror. Also, no apparent burrs,

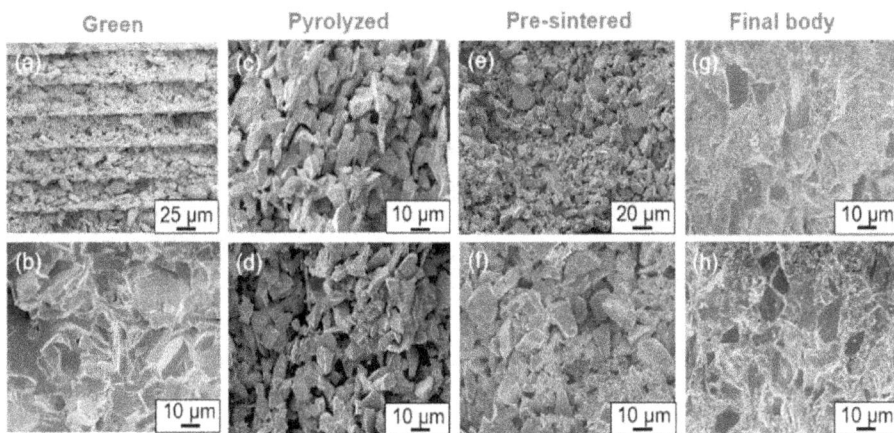

Figure 2.24. SEM images of the cross-section (top photos) and surface (bottom photos) of the SiC body at different post-processing stages: (a) and (b) green body; (c) and (d) pyrolysed body; (e) and (f) pre-sintered body; (g) and (h) final body. Reprinted from [128], Copyright (2020), with permission from Elsevier.

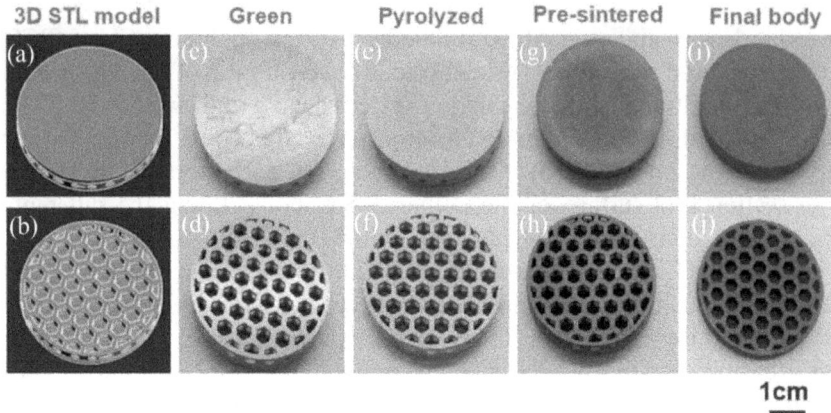

Figure 2.25. CAD images of the SiC ceramic optical mirror: (a) top part, (b) support structure. Photographs of the mirrors during different processing stages: (c) and (d) green body; (e) and (f) pyrolysed body; (g) and (h) pre-sintered body; (i) and (j) final body. Reprinted from [128], Copyright (2020), with permission from Elsevier.

cracks, porous structures, or breaks were detected. Furthermore, it was verified that the post-processing methods did not damage the surface roughness of the mirror and/or its structural integrity. Regarding the weight of the mirror, the authors verified that the mass of the manufactured mirror was reduced by 40.9% compared to the weight of a bulk ceramic mirror with the same dimensions, thus confirming the advantages of the AM SiC ceramic lightweight mirror for space applications.

2.3 Optomechanical components

Nowadays, traditional manufacturing methods used to fabricate precision optomechanical components through CNC machining require both time and complexity. This impacts the cost of the final part, restricting its widespread use and its associated benefits. Additionally, for a part to be suitable for a specific target application, it must possess certain properties, such as density, thermal resistance, expansion, flexibility, ease of processing, and an affordable cost. As our society evolves, more advanced optical systems are required; examples can be found in applications such as aerospace, astronomy, and precision instrumentation. AM allows printing in a variety of materials, enabling the exploration of different properties. Additionally, its resolution, print flexibility, and efficiency allow the design, fabrication, and testing of parts in an easy and cost-effective manner. This shift from traditional manufacturing accelerates both prototyping and production, enabling research, development, and innovation. It can also be an affordable way to produce cheap and practical components for teaching optics [129]. Through this manufacturing paradigm shift, researchers, engineers, teachers, and others can access various optical components that would otherwise require time to acquire due to lengthy delivery processes, incur high costs due to traditional fabrication processes, or have limited design freedom or properties.

2.3.1 Integrating spheres

Integrating spheres (ISs) are optical devices used to measure the luminous flux and calibrate devices that respond to the radiation spectrum. Examples of their use include the measurement of luminous flux, as in the case of the measurement of the total light output from lamps, LEDs, and other light sources; measuring diffuse reflectance, namely to characterise the reflective properties of surfaces, including textiles, paint, etc; and the generation of uniform light sources for calibration and imaging applications.

An IS is a spherical structure with a hollow cavity and two windows oriented 90° apart from each other. One of these windows works as the input window, and the other works as the output window, where the detector is located. It contains a small sample port area (less than 4% of the total area), and a baffle is also used to prevent direct radiation reflection from the sample to the detector. The inner surface of the IS is typically prepared with a uniform and highly diffuse reflective coating that does not exhibit fluorescence. Thus, when light enters the input window, it is reflected isotropically over the whole internal surface of the IS. This enables the device to measure optical power independently of the beam shape or entrance angle. An example of an IS is shown in figure 2.26.

ISs are now widely available and can be found in various optical equipment stores. Despite their ease of access and availability, the economic factor may limit their acquisition. The cost scales with the size of the IS, and values reaching several thousand euros are not uncommon. This is particularly relevant for low-income countries, newly formed research groups, hobbyists, and didactic instrumentation in classrooms, which cannot afford to have this device. Because of this, different authors have already dedicated their work to the development of affordable ISs [130–132]. For this purpose, FDM printers have been the common choice. An example is the study by Serrano *et al* [131], where an IS was 3D printed using polylactic acid (PLA) and later characterised in terms of its power intensity

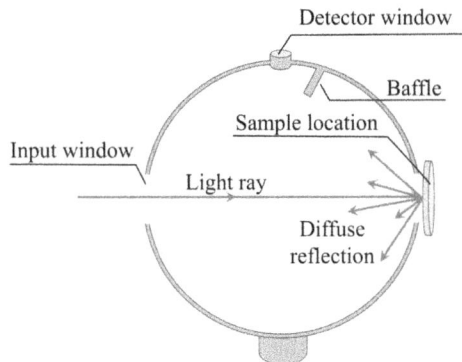

Figure 2.26. Schematic of an IS. A light ray travelling from left to right enters the input port and illuminates the sample. After multiple scattering reflections, distributed equally at the inner surface of the IS, the light is collected at the detector window, preserving power but destroying the spatial information.

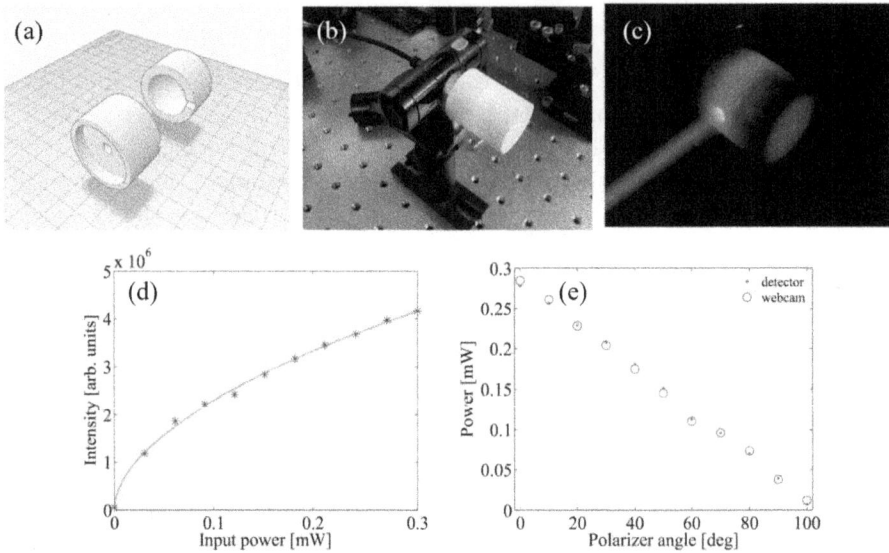

Figure 2.27. (a) CAD design of the IS split into two halves for visualisation purposes (this division was not taken into account during the printing process). (b) A 3D-printed IS is coupled to a webcam. (c) IS operation in a darkroom. (d) Calibration of the camera intensity distribution vs. input power (using a power meter). (e) Performance evaluation after the calibration. The input power is changed by rotating one linear polariser in front of another. Reproduced from [131]. CC BY 4.0.

measurement, utilising a low-cost webcam connected to its output window. The results of their study are compiled in figure 2.27.

To collect the intensity distribution from the camera in [131], the power intensity histogram of a certain pixel area was considered. Later, for calibration purposes, this intensity was plotted against the input power of a He–Ne laser, measured through a power meter. The results presented by the authors (see figure 2.27(d)) showed predictable nonlinear behaviour. Later, using the previously obtained calibration curve, the authors demonstrated the performance of the IS coupled to the webcam for different laser input powers (two linear polarisers were used to change the optical power by altering the angle between them), as shown in figure 2.27(e). The authors of this study developed a low-cost IS device with measurement errors close to 2%.

Despite the preliminary results on the topic of ISs, the study reported in [131] did not reveal information regarding the surface roughness and/or reflectance of the inner surface region. Since the IS was printed as a whole, it is assumed that no coating was used and that the roughness associated with the layering effect was present on the inner wall of the sphere, which could introduce several concerns related to its diffuse reflectance and its practical implementation in measurement and/or characterisation scenarios.

Tomes *et al* [132] also pioneered the development of ISs. In contrast to the study by Serrano *et al*, they 3D printed the IS in two separate halves. This allowed them to access the interior of the sphere for surface polishing, specifically to remove the printing layer effect and also to apply a surface coating to achieve diffuse reflectance.

To do so, they first sanded the surface with P120 grit sandpaper, then added a surface filler, followed by a second smooth polishing with P320 grit sandpaper. Finally, the surface coating was applied by spraying on several coats of a homemade reflective coating based on 99% purity barium sulphate (BaSO$_4$) powder mixed with 1% (by weight) polyvinyl acetate, achieving a reflectivity of around 90%–94%. The measurement was made by placing an optical source at the IS entrance window and collecting the signal at the output window using a fibre coupled to a charge-coupled device (CCD) spectrometer. After system calibration, the authors were able to determine the spectrum and intensity of state-of-the-art photoluminescent materials. In this way, they were able to estimate the photoluminescent quantum yield (the ratio of the number of photons emitted by the material to the number of photons absorbed). Their estimations were very close to those found in the literature, indicating that the developed 3D-printed IS was efficient in solving the problem related to light collection from samples with anisotropic angular emission.

In addition to diffuse reflectance, ISs can also measure diffuse transmittance. By doing so, and through the use of reference algorithms, such as inverse adding doubling (IAD), it is possible to measure optical properties such as average absorption (μ_a), the reduced scattering coefficient (μ'_s), and the anisotropy factor (g) in highly turbid samples [133]. Despite the preliminary studies described in [130] and [132] related to the development of ISs, information concerning the validation of the coating and reflectance of the proposed IS was limited. Because of that, Junior *et al* [130] decided to construct and fully characterise a 3D-printed IS. To do so, their fabrication followed the same procedure described earlier by Tomes *et al*. For the coating, pure BaSO$_4$ powder and a mixture of BaSO$_4$ and white paint were used. The characterisation was performed using two accurate optical polyurethane phantoms, allowing them to measure μ_a and μ'_s coefficients. A CAD representation and the 3D-printed halves of the 150 mm IS developed in [130] can be seen in figure 2.28.

The results reported in [130] and presented in figure 2.28(c) revealed that the reflectance of the BaSO$_4$ mixtures was similar to but lower than that of pure BaSO$_4$. Using the results presented in this figure for pure BaSO$_4$, the authors were able to compute the average standard coating reflectance ($r_{coating} = 0.97$) and the average wall reflectance ($r_w = 0.90$). These parameters were later used as correction factors in the IAD algorithm, allowing them to compute the μ_a and μ'_s coefficients for two different polyurethane phantoms, as depicted in figure 2.28(d), reaching values very close to the reference ones at 630 nm. This validated the proposed IS and demonstrated that the developed IS can successfully measure the absorption and reduced scattering coefficients in samples with both high and low absorption.

2.3.2 Optomechanical components

Optomechanical components are crucial in various optical systems, including laser cutting and precision manufacturing in the semiconductor industry, as well as in optical instruments such as microscopes, telescopes, and laser systems. They ensure that optical elements, such as lenses, mirrors, and filters, are properly located to

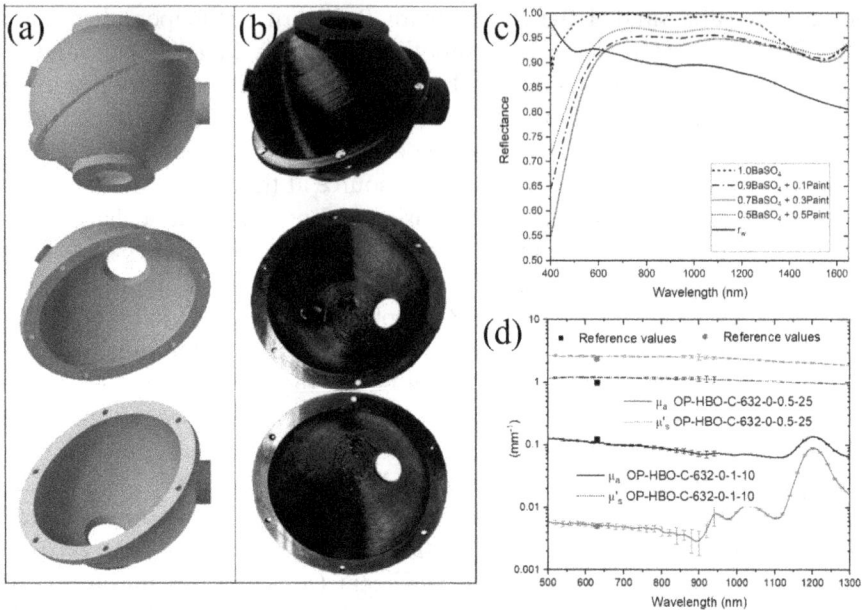

Figure 2.28. (a) CAD image and (b) 3D-printed parts of an IS produced by Junior *et al.* (c) Reflectance of the crafted coating compared to commercial PTFE and wall reflectance r_w of the sphere versus wavelength. (d) μ_a and μ'_s coefficients measured for two polyurethane phantoms through the developed IS and compared to a reference value at 630 nm [130] Taylor & Francis Ltd. http://tandfonline.com.

perform their intended functions. To achieve this, they must precisely and repeatedly adjust the positioning or angular orientation of optical elements, such as mirrors, lenses, or prisms, relative to the optical beam in an optical experiment. Furthermore, they need to guarantee stability to prevent changes in the optical path.

Optomechanical components can be as simple as mounts, posts, breadboards, and holders, capable of holding optical components in place and doing rough alignments, or more advanced, such as kinematic mounts (KMs) and translation stages (TSs), which allow for precise and repeatable adjustments of the angle or the position of an optical component in 3D space. KMs typically consist of three contact points, each made of ball bearings. These are placed such that they allow for pitch and yaw rotation and sometimes axial translation of the optical component (if properly designed for that). To perform these movements, the contact points have adjustment screws that allow tilting of the optic on a specific axis. The key attributes of these specialised holders are their stability, precision, and repeatability. On the other hand, TSs allow the movement of an optical element along a linear axis or in multiple axes when combined with multiple TSs. The most straightforward drive mechanism of these stages relies on the conversion of rotary motion to linear motion through a ball screw actuator.

Simple or complex, optomechanical components are widespread and easily found in optical laboratory stores. Their cost varies from tens to hundreds of euros, depending on the precision and size needed. On the other hand, low-cost, off-the-shelf 3D printers provide an accessible alternative for producing these components. This can eliminate the time lag between the purchase and delivery of a specific component, enabling research to progress without delay and, if necessary, the development and testing of a particular functionality in a timely manner. In addition, low-income laboratories, professors, and graduate students can benefit from creating their own optical components, allowing them to support projects and teaching activities. This has the potential to spread the use of optical technologies to all.

Serrano *et al* have been pioneers in developing optomechanical components through 3D printing. In one of their studies [131], they demonstrated the development and characterisation of a two-axis mirror KM and a one-axis TS. For this purpose, the authors employed an FDM 3D printer and the thermoplastic PLA as the extruded material to produce the different parts of the components. The CAD images, 3D-printed parts, and the assembled optomechanical components can be seen in figures 2.29(a)–(c) for the mirror KM and in figures 2.29(d)–(f) for the TS, respectively.

The assembly of the KM shown in [131] and presented in figure 2.29(c) required the use of two extension springs to secure the two 3D-printed parts close together while retaining enough freedom to allow the motion of one part over the other. The tilt is achieved with the help of adjustment screws that can freely rotate on nuts fixed to the top 3D-printed part of the KM. These screws had tight rounded nuts at the terminals to allow their rotation, with low friction, at the grooved locations of the second KM 3D-printed part. A steel sphere was used to allow free rotation in each of the two axes.

The TS presented by Serrano *et al* [131] and shown in figure 2.29(f) is composed of two 3D-printed parts: one serving as the static platform, while the other is allowed to move by approximately 1 cm. To achieve this, they used two steel rods fixed to the static 3D-printed platform. The top platform had two through holes that served as slide guides for the steel rods. The movement (backward and forward) is achieved using an

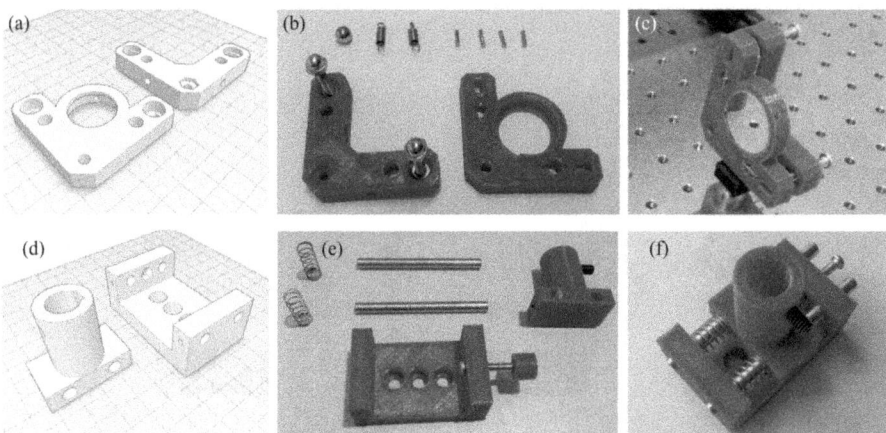

Figure 2.29. Images related to the (a–c) mirror KM and (d–f) TS. Panels (a) and (d) show CAD drawings of the parts, (b) and (e) show photos of the 3D-printed parts, and (c) and (f) show photos of the assembled components. Reproduced from [131]. CC BY 4.0.

M4 screw that rotates in the clockwise and counterclockwise directions, respectively, on a nut fixed to the static 3D-printed platform. For the backward motion, two compression springs pull the top movable platform along the guiding steel rods.

To fully characterise the fabricated optomechanical components, Serrano *et al* [131] evaluated the production cost, time from requisition to availability, and performance in operation.

Regarding the production cost, the authors used less than 30 g of PLA to print the two parts of each optomechanical component. Considering PLA market pricing of approximately $20 kg^{-1}, this leads to a material cost of just a few cents per part. Assuming a price of 25 cents for each of the parts involved in each of the optomechanical components (i.e. springs, metal rods, nuts, bolts, etc.), the estimated budget for each element can be as low as a few dollars. In comparison to marketed optomechanical components, the 3D-printed components can achieve costs that are 10 and 30 times lower than those of commercially available KMs and TSs, respectively. It is worth mentioning that the prices of these components increase with the dimensions of the optics, making 3D-printed mounts even more attractive.

Concerning the fabrication time, the 3D-printed parts took less than three hours to print, and some labour was required for assembly. This is a great opportunity, considering the lead time necessary for product delivery, which can sometimes take several days or even weeks in some regions of the globe. This print-on-demand approach also presents an opportunity to accelerate research production, as setups can be easily tested, characterised, and modified within a short period.

For the performance evaluation, the authors created dedicated characterisation setups and compared the results obtained using the 3D-printed components with those obtained using commercial components [131]. Both setups utilised a Gaussian beam 1 mm in diameter produced by a He–Ne laser, two commercial KMs, a webcam, two irises, and the 3D-printed optomechanical component (with an additional beam splitter in the case of the TS characterisation setup). The setups implemented for the KM and the TS can be seen in figures 2.30(a) and (b), respectively.

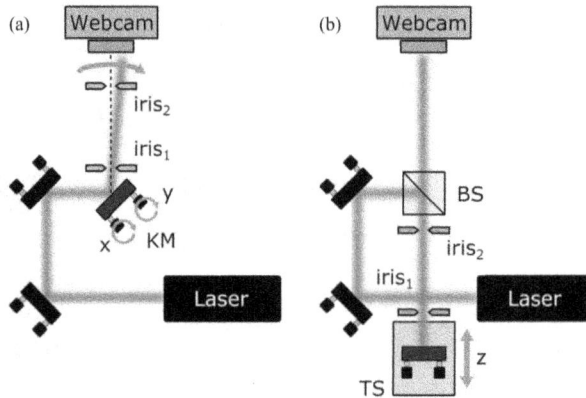

Figure 2.30. Schematic of the setups used for the characterisation of the (a) KM, (b) TS. Reproduced from [131]. CC BY 4.0.

Regarding the KM, the tests were performed ten times and consisted of tilting a mirror secured to the KM holder, horizontally and vertically, through the rotation of the X and Y screw knobs of the KM, respectively, and observing the beam deviation on the camera sitting 80 cm from the mirror [131]. For the TS characterisation, two sets of measurements were taken by displacing the position of a mirror sitting on top of the mobile part of the TS in 2 mm increments. The results for the KM are shown in figures 2.31(a) and (b), while those for the TS are shown in figure 2.31(c).

From the centroid position results obtained for both the commercial KM and the printed one, it can be observed that there is good agreement between them. The beam position obtained for the 3D-printed KM followed predictable values, with minor deviations observed, demonstrating its suitability for deployment in optical systems. Regarding the TS, the results were less satisfactory but still acceptable within a displacement range of 0–6 mm. For ranges above this, the compression springs imposed too much strain on the movable platform, which induced possible unwanted deviations, leading to worse results than those achieved with the commercial TS.

Optomechanical components sometimes include automation to enhance user interaction in a given application. In this regard, the team of Salazar-Serrano showed the capability to do this in the KM shown in figure 2.29(c) by integrating Arduino-controllable stepper motors to control the rotation of the adjustment screws of the KM and thus perform beam alignment in an automated fashion [134].

The low-cost opportunities presented by FDM and SLA 3D-printing technologies enable the sharing and dissemination of knowledge to a broad audience. Gunderson *et al* did pioneering work in designing and freely sharing several optomechanical components capable of being custom edited and 3D printed [135]. Examples of these parts are shown in figure 2.32.

One important thing to note in figure 2.32 is the general usage of brass heat-set inserts (shown in yellow) found in some 3D-printed components. These are used to add threaded mounting points, allowing for easy assembly of the parts and increasing their lifespan, which would otherwise be shortened due to the poor mechanical strength of polymers. While the embedment of these inserts is easily achieved in thermoplastic materials (extruded from FDM printers) through the use of a hot soldering iron, the polymer materials used in SLA printers are thermoset;

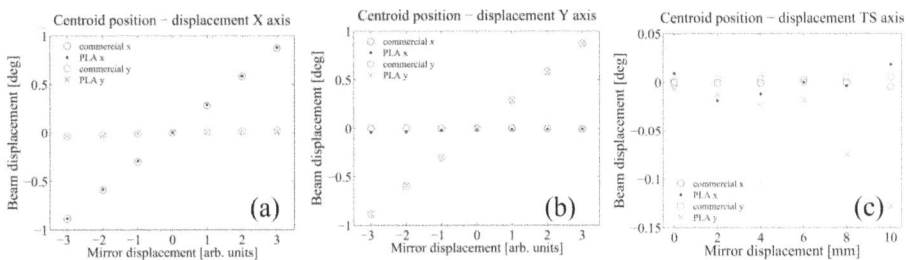

Figure 2.31. Centroid beam positions for the commercial and 3D-printed KMs when each of the knobs (a) X and (b) Y is adjusted. (c) Centroid beam position for the commercial and 3D-printed TSs as a function of the displacement imposed on the movable part of the TS. Reproduced from [131]. CC BY 4.0.

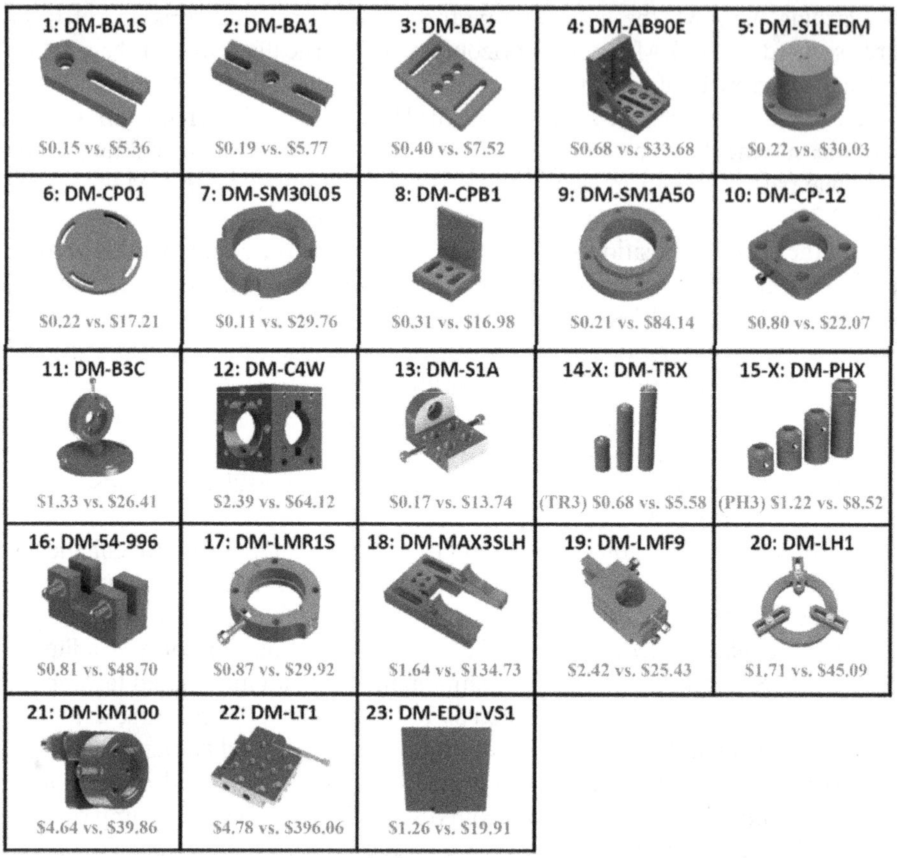

1: DM-BA1S	2: DM-BA1	3: DM-BA2	4: DM-AB90E	5: DM-S1LEDM
$0.15 vs. $5.36	$0.19 vs. $5.77	$0.40 vs. $7.52	$0.68 vs. $33.68	$0.22 vs. $30.03
6: DM-CP01	7: DM-SM30L05	8: DM-CPB1	9: DM-SM1A50	10: DM-CP-12
$0.22 vs. $17.21	$0.11 vs. $29.76	$0.31 vs. $16.98	$0.21 vs. $84.14	$0.80 vs. $22.07
11: DM-B3C	12: DM-C4W	13: DM-S1A	14-X: DM-TRX	15-X: DM-PHX
$1.33 vs. $26.41	$2.39 vs. $64.12	$0.17 vs. $13.74	(TR3) $0.68 vs. $5.58	(PH3) $1.22 vs. $8.52
16: DM-54-996	17: DM-LMR1S	18: DM-MAX3SLH	19: DM-LMF9	20: DM-LH1
$0.81 vs. $48.70	$0.87 vs. $29.92	$1.64 vs. $134.73	$2.42 vs. $25.43	$1.71 vs. $45.09
21: DM-KM100	22: DM-LT1	23: DM-EDU-VS1		
$4.64 vs. $39.86	$4.78 vs. $396.06	$1.26 vs. $19.91		

Figure 2.32. Images of freely available CAD optomechanical components for 3D printing. A price comparison is also included for each 3D-printed component (the price of PLA is assumed to be $25.00 kg^{-1}) and its corresponding commercially available product from Thorlabs Inc. (prices from 2020). Adapted with permission from [135]. Copyright (2020) American Chemical Society.

therefore, these inserts can only be added with the help of some adhesive applied between the insert and the cavity intended to contain it.

Gunderson *et al* [135] also provided a general cost comparison between the costs of the 3D-printed components and those commercially available from Thorlabs Inc., as detailed under each component shown in figure 2.32. The conclusions are clear: cost reductions for the 3D-printed components compared to commercial parts average 94%, and in some cases, a direct comparison results in a 99% reduction. The results demonstrate the potential for opening the fields of optics and photonics, typically restricted to optical research laboratories, to a broader audience.

2.4 Setups built from 3D-printed parts

The ability to 3D print optomechanical components can support a specific objective in an optical system, allowing creativity to be incorporated into the setup design or even addressing an urgent need that would otherwise take time to resolve. Given these clear opportunities, the possibility of exclusively using 3D-printed optome-chanical components in scientific setups or even scientific instruments is well documented.

Regarding the toolbox of components presented by Gunderson *et al* [135], described earlier in figure 2.32, the authors demonstrated a Michelson interferom-eter. Functional toolboxes in the form of unit box-like approaches have also been shared by other authors [136–138], allowing, for instance, the study of the Michelson interferometer or its use in Raman spectroscopy [138]. Three-dimensional printed setups used for microscopy have also been well documented in the literature [135, 136, 139, 140]. Specifically considering the toolbox of components reported in [135] and shown in figure 2.32, the authors used seventeen of these optomechanical components (i.e. parts numbered 4–13, 14–1, 15–1, 17, 18, and 22) to build an optical microscope. In addition, the authors also used standard optical and electrical components to support the proper operation of the microscope, such as the use of an LED as a source of light (which was passed through a condenser lens to illuminate the specimen), a 10X objective lens to collect transmitted light from the specimen, and two lenses and one mirror to guide the light to a CCD camera. A schematic of the microscope, its CAD assembly, and a photo of the instrument are shown in figure 2.33.

To test the microscope shown in figure 2.33(c), the authors collected images from a calibration ruler slide and from a biological sample; specifically, a cross-section of

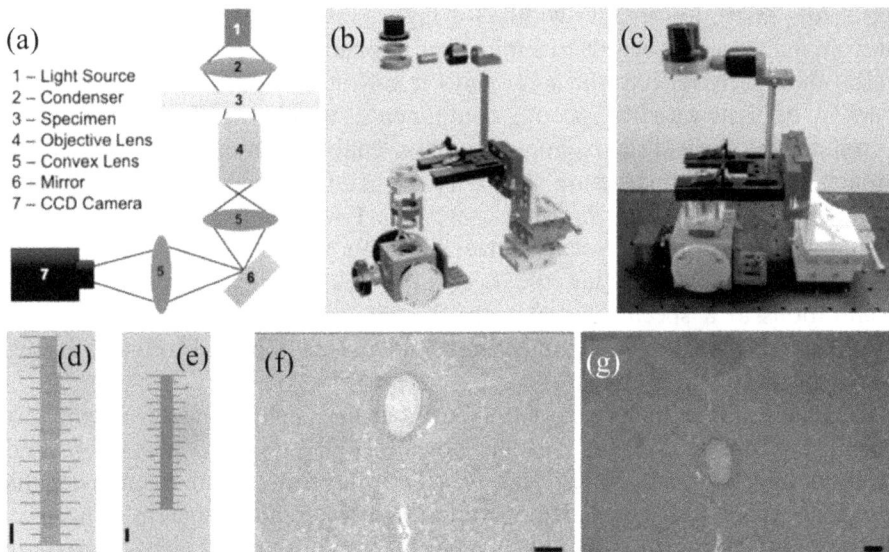

Figure 2.33. Construction of an inverted brightfield microscope that includes 3D-printed optomechanical components: (a) schematic drawing of the microscope; (b) pre-assembled CAD optomechanical components; (c) microscope fully assembled with 3D-printed parts. Images collected using: (d), (f) the home-built microscope and (e), (g) the commercial microscope. (The scale bar is 100 μm.) Adapted with permission from [135]. Copyright (2020) American Chemical Society.

a rabbit's spinal cord. The results were later compared to those obtained using a commercial brightfield microscope (EVOS FL from Thermo Fisher Scientific), which had the same objective magnification. The results are depicted in figures 2.33 (d) and (f) for the homemade microscope and in figures 2.33(e) and (g) for the commercial one. The images showed a higher magnification for the home-built microscope (probably due to the set of lenses used after the objective lens) and revealed more contrast, which was probably associated with the camera used in the home-built microscope. The results successfully demonstrated that an in-house microscope can be fully designed and assembled using 3D-printed parts, yielding comparable or even better results than commercial devices at a fraction of the cost.

Another example of 3D-printed setups can be found in [141], where the authors developed a low-cost UV–Vis spectrometer aided by a smartphone. The idea behind this study was to make absorbance spectroscopy more accessible and open it to a broader audience. To illustrate this concept, the authors effectively integrated optical components into 3D-printed parts in an intuitive and accessible manner. A schematic and a photo of the 3D-printed phone-assisted spectrometer can be seen in figures 2.34(a) and (b), respectively.

The schematic of the spectrometer presented in figure 2.34 was designed by considering the regular components of an absorbance spectrometer. Its principle of operation is that light from an external source, such as an LED or lamp, is passed through a 3D-printed entrance aperture slit, which shapes the light beam and controls its intensity before it is projected onto a sample contained in a cuvette. This aperture is interchangeable, allowing the user to control and test the principle of spectral resolution, i.e. smaller widths for higher resolutions and vice versa. Then, light from the sample proceeds to a mirror positioned at 45° relative to the incoming light beam, which reflects the light onto a diffraction grating. This allows the dispersion of light into its spectral components, which are then captured by a detector, in this case, a smartphone's camera, enabling an easy and cost-effective approach. The housing structure is entirely 3D printed, and all the other necessary parts are assembled in the housing. After pixel–wavelength calibration using a known peak-intensity fluorescent source, the authors demonstrated the spectrometer's capabilities by verifying the Beer–Lambert law for solutions of varying concentrations at a specific wavelength, thereby showcasing the potential use of this spectrometer for laboratory measurements and its potential to increase shared knowledge [141].

Other interesting systems reported in the literature include the development of a 3D-printed fluorometer [142], an ellipsometer [143], a polarimeter [144], and a Michelson interferometer [135], among others. Overall, all the optomechanical components and systems have demonstrated significant advantages in terms of cost reduction, fabrication time, simplicity, and customisation capabilities. Despite this, it is essential to note that the performance of the final 3D-printed assembled part is always dependent on the skills of the person involved in the fabrication, the parts used to support the mechanism (springs, rods, bearings, etc.) and also on the material used for the 3D printing. While consumer-grade FDM and SLA 3D printers offer opportunities in terms of cost and production time, they can still create

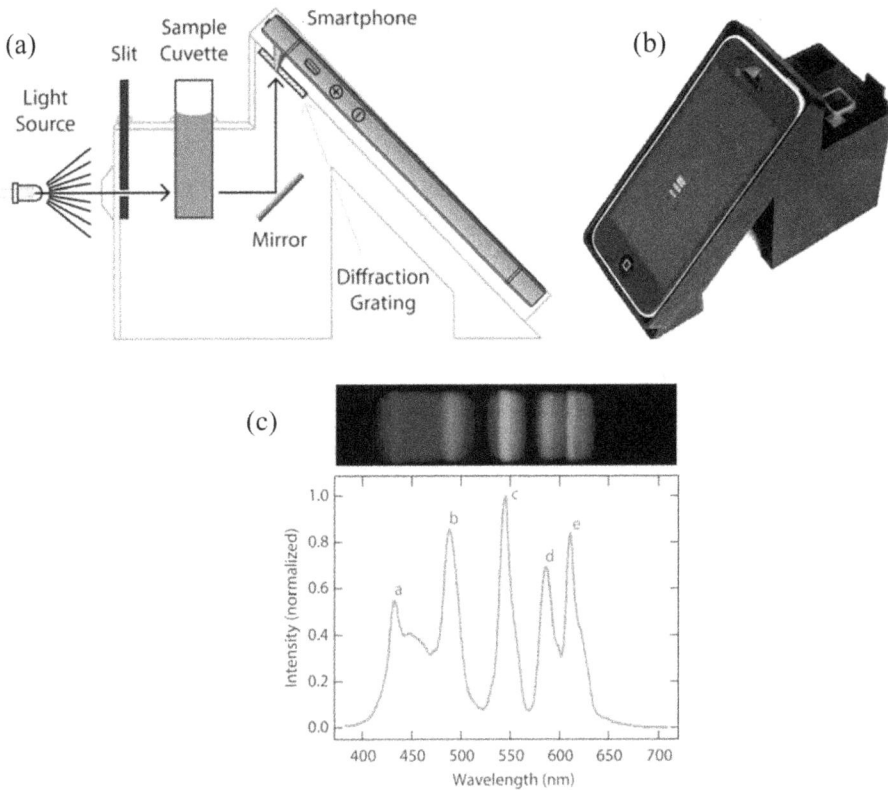

Figure 2.34. (a) Schematic of the absorbance spectrometer presented in [141], where light produced by a source at the left is filtered through a removable slit, passes through the sample, reflects off a mirror, and diffracts on a dispersion grating positioned near a smartphone camera used to capture the diffracted image pattern. (b) 3D-printed spectrometer fully assembled. (c) Light from a fluorescent source with a known peak-wavelength intensity is diffracted after passing through the grating. A photo is captured by the smartphone camera (top image) and processed to show the intensity vs. wavelength after pixel–wavelength conversion. Reprinted with permission from [141]. Copyright (2015) American Chemical Society.

issues regarding mechanical and thermal stability. For instance, the thermoplastic polymers used in FDM printers can deform easily at low temperatures or under minimal strain. Additionally, the thermosetting polymers used by SLA printers are brittle and can easily break under high loads or burn at high temperatures. Thus, despite the possibility of creating inexpensive, fast, and on-demand optomechanical components, a proper understanding of their final application is required. Otherwise, the overall system or equipment performance can be compromised.

2.5 Optical fibre structures aided by 3D printing

Optical fibre gratings are important passive components found in the telecommunications and sensing fields. Their capability to manipulate light propagation inside optical fibres is unique, allowing them to act as spectral filters and to present multiplexing capabilities. This manipulation occurs due to the variation in refractive index induced by the grating along the fibre length.

Fibre gratings can be grouped into two main categories: short-period gratings, also known as fibre Bragg gratings (FBGs), and long-period gratings (LPGs). The difference between them lies in their period: the period of the former is on the order of hundreds of nanometres, while the period of the latter is on the order of several tens of micrometres to a millimetre [145]. In these periodic refractive index structures, the fundamental co-propagating core mode can couple to the counter-propagating core mode, as is the case with FBGs, and to the co-propagating cladding modes in the case of LPGs. For FBGs, the transmission spectrum presents a localised attenuation band for a wavelength that satisfies the Bragg condition. Regarding LPGs, the transmission spectrum has specific attenuation bands corresponding to the coupling of the core mode to a particular cladding mode. The location where these resonance bands occur is described by the following expression [146]:

$$\lambda_i = (n_e - n_i)\Lambda \qquad (2.4)$$

where n_e is the effective refractive index of the core mode, n_i is the effective refractive index of the ith cladding mode, and Λ is the period of the refractive index modulation or, in other words, the period of the grating.

LPGs possess important features, including low out-of-band losses, good signal-to-noise ratios, versatility (they can be manufactured with different periods to tune the resonant wavelength), and high sensitivity to several parameters. Due to these advantages, there have been reports of the use of these fibre devices for various applications, including gain flattening in erbium-doped fibre amplifiers, gain equal-isers [147], comb filters [148], mode converters [149], and sensors [150], among others [151]. Regarding sensor applications, LPGs have been reported to measure several parameters, including torsion, curvature, strain, temperature, and humidity, among others.

2.5.1 Three-dimensional printed amplitude masks for UV-LPG inscription

Currently, there are two main types of LPGs: those that are permanently inscribed in the fibre and those that are mechanically induced. Regarding the permanently inscribed type, several fabrication methods have already been reported, such as irradiation by CO_2 laser [152], UV laser [153], infrared femtosecond laser [154], ion beam [155], and electric arc discharge [156]. Due to the widespread use of deep UV lasers in lithography processes and the extensive deployment of UV-inscribed FBGs, UV laser methods have been among the most frequently reported methods in the literature. The UV inscription of LPGs can be achieved in two ways: either through a point-by-point approach, where a positioner moves the fibre or a mirror to irradiate the fibre in a periodic fashion, or by using an amplitude mask that already contains the periodic pattern to be inscribed into the fibre. While the point-by-point method offers flexibility in defining the grating period, it is more time-consuming and requires more components. On the other hand, amplitude masks allow for a more straightforward setup and operation and are more suited for mass production.

However, in contrast to FBGs, where it is easy to find phase mask suppliers, the same is not true for LPGs. Thus, amplitude masks have been fabricated by subtractive methods through precision laser milling systems capable of drilling patterns of holes in thin metallic sheets. Such an approach is very promising, but the cost, labour, and time required to make multiple metal masks with specific grating periods or profiles could be prohibitive. Moreover, metal masks are easy to oxidise and deform. AM methods offer a simple, cost-effective solution for producing these masks with any arbitrary period or profile, at a fraction of the cost of subtractive methods. Oliveira *et al* [157] considered these opportunities and demonstrated the feasibility of producing these masks using a consumer-grade DLP 3D printer. A proof of concept was described for nine amplitude masks, each consisting of 100 periods, with a thickness of 5 mm, a height of 16 mm, and periods ranging from 690 to 950 µm. A CAD image of one of the phase masks, along with its printed version, is shown in figure 2.35.

An LCD screen with a resolution of 47 µm was sufficient to reproduce the tiny periodic pattern of the mask. However, to be confident about the performance of the masks, the authors irradiated the masks with a focused UV beam and observed the resulting UV patterns on thermal paper. The results for three different mask periods, i.e. 690, 820, and 950 µm, are shown in figure 2.36 [157].

As can be seen in figure 2.36, the phase masks produced in [157] showed the capability to periodically imprint the UV pattern on the thermal paper, demonstrating the effectiveness of the printed masks. Slight discrepancies were, however, attributed to the orientation of the masks relative to the beam and to the resolution of the printer.

The LPG inscription capabilities of the 3D printed amplitude masks were later tested by scanning the UV laser beam through one of the fabricated amplitude masks in a photosensitive single-mode fibre (SMF). Examples of the transmission spectra obtained for different grating lengths in [157] can be seen in figure 2.37.

The results obtained for a grating with a 690 µm period revealed the appearance of three resonant bands in the 1550 nm region, showing greater strength as the grating length increased. The dip resonances reached maximum transmission losses of $\approx$18 dB for a grating length of $\approx$38 mm. The out-of-band losses reached low values (i.e. <0.2 dB), and the 3 dB bandwidth was $\approx$3 nm. In addition, the wavelengths of the spectral dips were easily controlled through the use of amplitude masks with different grating periods. Experimentally, the team demonstrated that the dip resonances could be linearly tuned over approximately 300 nm by varying

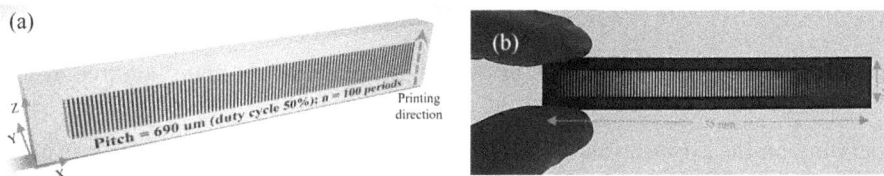

Figure 2.35. (a) CAD image of an amplitude mask and (b) its 3D printed version. Reproduced from [157].

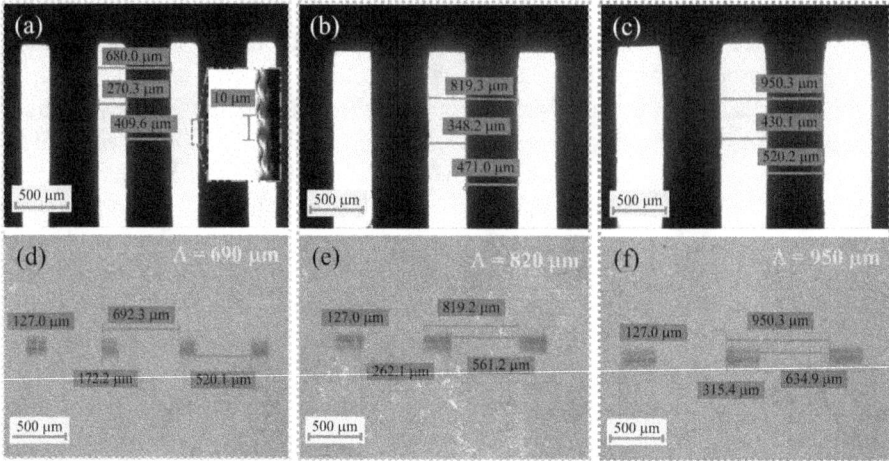

Figure 2.36. Microscope images of the amplitude masks (a–c) and the UV patterns imprinted on thermal paper using those masks (d–f). The images, from left to right, correspond to the following periods: 690, 820, and 950 μm, respectively. Reproduced from [157]. CC BY 4.0.

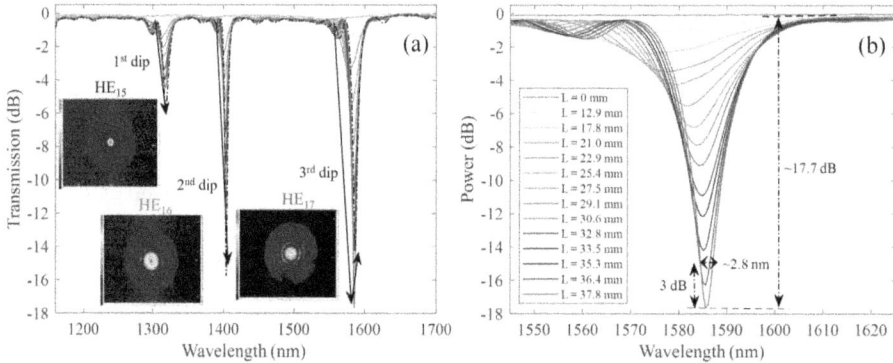

Figure 2.37. (a) Transmission spectra of UV-inscribed LPGs with different grating lengths obtained using 3D-printed amplitude masks with a 690 μm period. The inset near-field images correspond to the dip resonances. (b) Magnified view of the third LPG dip resonance for different grating lengths. Reproduced from [157]. CC BY 4.0.

the amplitude mask period from 690 to 950 nm. Overall, the results demonstrated the feasibility of using 3D-printed amplitude masks for the effective UV inscription of LPGs.

2.5.2 Mechanically induced LPGs produced using additively manufactured grooved plates

Depending on the application, LPGs may need to be tuned for both dip loss and dip wavelength after their inscription. Permanently inscribed LPGs, such as those explored in the previous paragraphs, can only be tuned by a few hundred picometres when subjected to external variables, such as strain or temperature. Tunability has been explored for the development of fibre sensor applications. The inscription of LPGs with different grating periods can shift the resonant wavelength to a certain region of the spectrum. However, after they have been inscribed, their dip resonant losses and wavelengths can barely be tuned.

Pressure-induced LPGs were the first type of grating to be reported in the literature [158]. They were created by pressing a periodic grooved structure onto an optical fibre (see the example in figure 2.38(a)). The resulting periodic strain distribution leads to periodic changes in the fibre's refractive index due to the photoelastic effect, as described by the following equation:

$$\Delta n = -n_0 p_e \varepsilon = -n_0 p_e \frac{F}{EA} \tag{2.5}$$

where n_0, p_e, ε, and F are the unperturbed refractive index, the effective photoelastic constant, the associated strain, and the applied force, respectively. Here, E and A are the effective Young's modulus and the radius of the region subjected to stress, respectively. As a result of this perturbed refractive index change, coupling occurs between the fundamental core mode and the forward-propagating cladding modes, as detailed in equation (2.4). This methodology offers two simple features: control of the coupling strength through the applied pressure and control of the resonant wavelength through the selection of the grooved plate. Again, these grooved plates are produced through subtractive methods, and thus, AM has provided an opportunity to readily produce these structures in a simple, fast, reliable, and cost-effective manner. Thus, Iezzi et al [159] were the first to report the 3D printing of grooved plates for the creation of pressure-induced LPGs. The structures were fabricated with an FDM printer that had a coarse resolution of 200 μm. As a result of this poor resolution, the team reported spectral resonances with high out-of-band losses (i.e. 5 dB) and poor spectral signatures. However, a few years later, Oliveira et al reported the use of amplitude masks produced using a DLP 3D printer (see figure 2.35), featuring a higher resolution (i.e. 47 μm) [157]. This allowed the team to obtain sharper dip resonances with low out-of-band losses (i.e. 0.6 dB). The results were reported for different grating periods, demonstrating the capability to spectrally shift the dip resonances. The results of this characterisation are shown in figure 2.38.

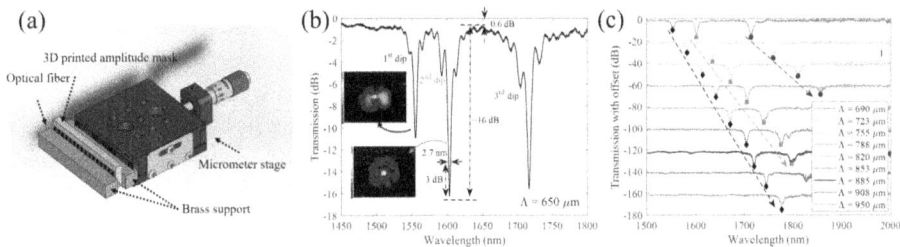

Figure 2.38. (a) Setup used to mechanically induce an LPG through pressure via a 3D-printed amplitude mask. (b) Transmission spectra for LPGs mechanically induced by applying pressure to an optical fibre through an amplitude mask. Transmission spectra were obtained for 3D-printed amplitude mask periods of (a) 690 μm and (c) 690–950 μm. Parts (b) and (c) reproduced from [157]. CC BY 4.0.

The spectral tuning of the dip resonances presented in figure 2.38 was easily achieved by adjusting the grooved periodic plate, resulting in the spectral wavelength shift shown in figure 2.38. Despite the opportunity to spectrally shift the spectra through the replacement of the grooved plate, its practicality is unappealing. A clever idea came from the study presented in [160], which reported a fully 3D printing system capable of rotating the fibre relative to the grooved plate, allowing the period induced in the fibre to be tuned by $\Lambda_{\text{tilt}} = \Lambda/\cos(\theta_t)$, where Λ is the period on the grooved plate and θ_t is the tilt angle.

From a sensor point of view, pressure-induced LPGs can themselves act as lateral pressure sensors, since the transmission dip power (T) follows a cosine-squared relationship of the form $T = \cos^2(k_{co-cl}^{ac}l)$, where k denotes the coupling strength of the core mode to a particular cladding mode, and l denotes the grating length. This allows the establishment of a relationship between T and the applied pressure, and thus enables sensor functionality to be achieved by measuring the transmission dip power. Turning to a different scenario, where the applied pressure is kept constant, the coupling strength should also remain constant; thus, no changes are observed in T. However, this is only true when the environmental temperature remains constant. If the temperature increases or decreases, materials expand or contract, respectively. Thus, the material applying periodic pressure to the fibre responds to temperature changes by increasing or decreasing the applied pressure, or even altering the grating period. This, in turn, affects either the resonant wavelength or the optical power, which can therefore be used to establish a relationship that can be used to indirectly measure the temperature by monitoring the transmission dip power. This principle was explored in reference [161], where a 3D-printed grooved plate kept a fibre under constant pressure and was used to measure the temperature.

The use of 3D-printed grooved plates capable of generating mechanically induced LPGs has also been explored to measure other variables, such as shear strain (γ) and torsion (θ) [162]. To achieve this, Oliveira *et al* used the 3D-printed grooved plate shown in figure 2.35 and bonded the fibre to the periodic structure using UV radiation, resulting in the structure shown in figures 2.39(a) and (b).

When they subjected the assembled structure presented in [162] and shown in figure 2.39(a) to increments of shear strain by immobilising it at one corner and applying strain to the diametrically opposed corner, the authors observed an increase in spectral dip resonances (see figure 2.39(c)). The same observation was made in the torsion tests, wherein one of the terminals of the 3D-printed structure was fixed while the other was rotated (see figure 2.39(d)). A deeper investigation into the dip optical power of these resonances produced the graphs shown in figure 2.40.

The results shown in figure 2.40 clearly demonstrate the linear response of the dip power as a function of the applied force. These results clearly demonstrate the powerful capabilities of 3D printing in equipping optical fibres to measure complex external parameters, as in the cases of shear strain and torsion. In addition to these capabilities, the assembled structure had also demonstrated the ability to measure temperature [162], or even the capability to measure curvature, as was reported in subsequent research by the same group [163].

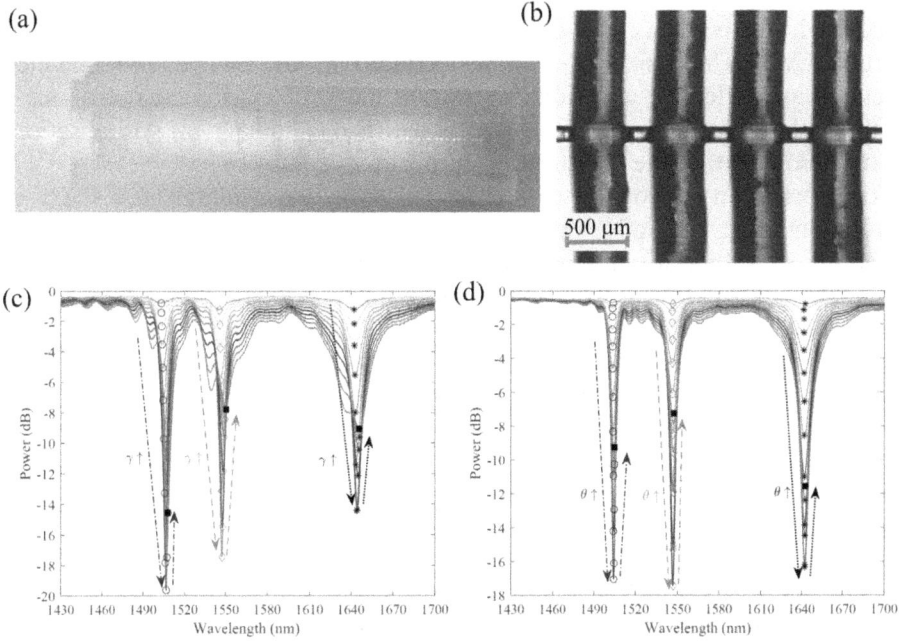

Figure 2.39. SMF bonded to the 3D-printed grooved plate (a). Magnified view of the bonded region (b). Optical spectra collected for different values of shear strain and torsion applied to the assembled 3D-printed structure. Adapted with permission from [162] © 2021 Optical Society of America.

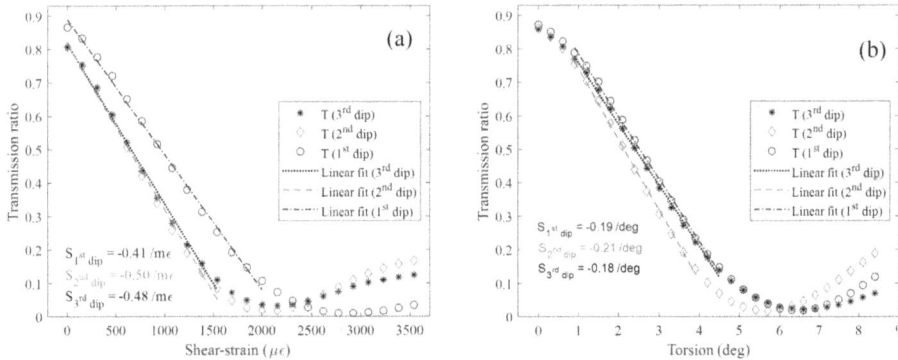

Figure 2.40. Transmission ratio as a function of shear strain (a) and torsion (b) for the three dip resonances that appear in figures 2.39(c) and (d), respectively. Reprinted with permission from [162] © 2021 Optical Society of America.

2.5.3 Corrugated LPGs assisted by 3D printing

Another type of mechanically induced LPG that has been reported in the literature is the corrugated LPG [164–166]. These are fibre structures that include periodic etched regions along the length of the fibre. The working principle behind these structures is the periodic change in the refractive index of the fibre that occurs when an external load, such as strain, is applied to the fibre terminals, as described by equation (2.5). Thus, since the strains induced in the etched and non-etched structures are different, a periodic refractive index change is created along the length of the fibre, and equation (2.4) is satisfied, allowing the visualisation of spectral resonances in the transmission spectrum.

The fabrication of corrugated LPGs has been achieved through the use of photolithographic methods, i.e. through the use of photoresists and etching procedures [164–166], which requires sophisticated and expensive equipment, along with qualified personnel. With this in mind, Valente *et al* [167] proposed the use of 3D-printed comb-like structures to assist in the fabrication of these periodically etched fibres. The idea behind their study was to periodically deposit droplets of resin along the length of the fibre to protect specific regions from the silica etching process. Their work consisted of the CAD of the comb-like structure, consisting of 200 μm diameter pillar structures periodically arranged in a rectangular support. The structure was printed by a DLP 3D printer with a resolution of 50 μm and is shown in figures 2.41(a)–(c).

After printing, the comb-like structure was placed on a setup, and the 'pillar' terminals were immersed in a photopolymerisable liquid resin. Small UV resin droplets adhered to each 'pillar' structure and were later brushed onto the fibre,

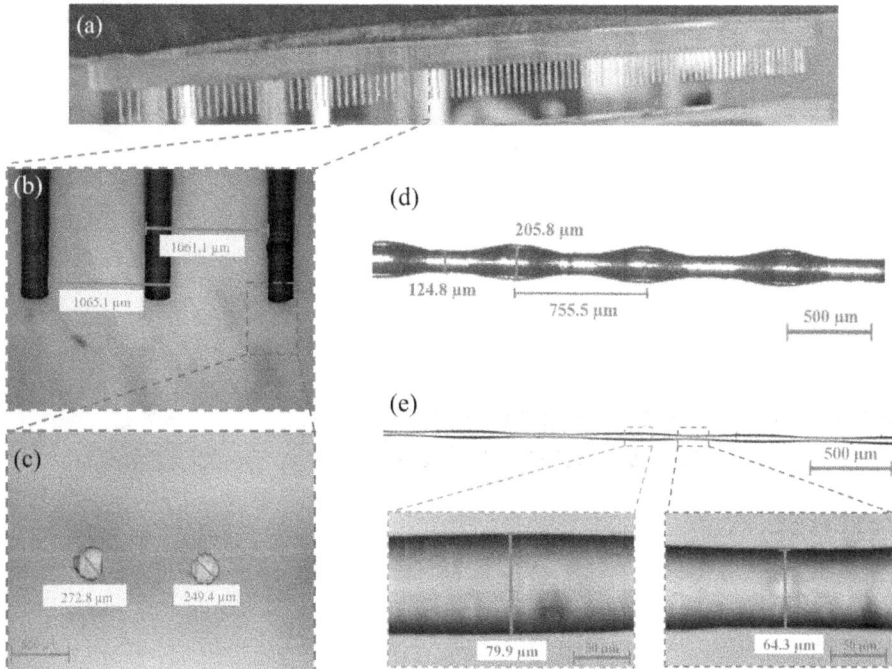

Figure 2.41. (a) Photo of the comb-like structure. Microscope images of the comb 'pillars' from the side (b) and the top (c). (d) SMF with added periodic droplets of polymerised UV resin. (e) SMF after the etching process. The insets show the diametric differences between the etched and non-etched regions. © [2024] IEEE. Reprinted, with permission, from [167].

leaving a periodic pattern of liquid resin that was later hardened by a UV lamp [167]. The result of the process is shown in figure 2.41(d). To create different droplet periods, the comb-like structure could be tilted relative to the fibre, exploring the same concept as described earlier in reference [160]. Finally, the post-processed fibre was immersed in hydrofluoric acid for ≈19 min, producing a periodically etched fibre structure as identified in figure 2.41(e).

The characterisation of the structure was performed using two parameters, namely strain and displacement. In the first case, strain was induced between the fibre terminals, and in the second case, the corrugated structure was put in a balloon-like shape, with one terminal fixed and the other displaced. The spectral results and the associated dip power and wavelength during the characterisation can be seen in figure 2.42 [167].

The results showed that the dip power resonance responded linearly to the imposed strain and displacement. The capability to measure displacement was demonstrated in a range of ≈6 cm, which showed that the sensor also had a linear wavelength response. Comparing the corrugated structure to those already reported in the literature, it was shown that the use of simple AM structures can reduce the cost and complexity of production, promoting a more widespread adoption of AM technology.

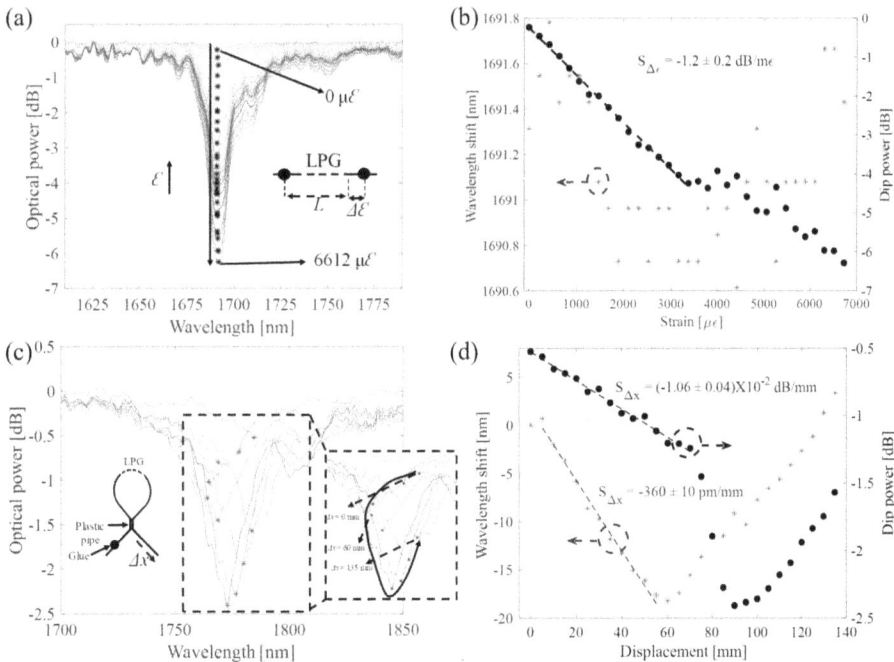

Figure 2.42. Transmission spectra (a), (c) and associated resonant wavelengths and dip optical powers (b), (d) for characterisations based on longitudinal strain (a), (b) and displacement (c), (d). © [2024] IEEE. Reprinted, with permission, from [167].

References

[1] Dick L 2012 High precision freeform polymer optics *Opt. Photon.* **7** 33–7

[2] Cogburn G, Mertus L and Symmons A 2010 Molding aspheric lenses for low-cost production versus diamond turned lenses *Infrared Technology and Applications* **XXXVI** (Bellingham, WA: SPIE) p. 766020

[3] Parks R E 1981 Overview of optical manufacturing methods *Contemporary Methods of Optical Fabrication* (Bellingham, WA: SPIE) pp. 2–12

[4] Baumer S 2010 *Handbook of Plastic Optics* (New York: Wiley)

[5] Schaub M P 2009 *The Design of Plastic Optical Systems* (Bellingham, WA: SPIE)

[6] Golini D, Jacobs S D, Kordonski W I and Dumas P 1997 Precision optics fabrication using magnetorheological finishing *Advanced Materials for Optics and Precision Structures: A Critical Review* (Bellingham, WA: SPIE) p 102890H

[7] Voelkel R 2012 Wafer-scale micro-optics fabrication *Adv. Opt. Technol.* **1** 135–50

[8] Gawedzinski J, Pawlowski M E and Tkaczyk T S 2017 Quantitative evaluation of performance of three-dimensional printed lenses *Opt. Eng.* **56** 084110

[9] Assefa B G, Pekkarinen M, Partanen H, Biskop J, Turunen J and Saarinen J 2019 Imaging-quality 3D-printed centimeter-scale lens *Opt. Express* **27** 12630–7

[10] Assefa B G *et al* 2019 Realizing freeform lenses using an optics 3D-printer for industrial based tailored irradiance distribution *OSA Contin.* **2** 690–702

[11] Assefa B G, Saastamoinen T, Biskop J, Kuittinen M, Turunen J and Saarinen J 2018 3D printed plano-freeform optics for non-coherent discontinuous beam shaping *Opt. Rev.* **25** 456–62

[12] Berglund G D and Tkaczyk T S 2019 Fabrication of optical components using a consumer-grade lithographic printer *Opt. Express* **27** 30405

[13] Gissibl T, Thiele S, Herkommer A and Giessen H 2016 Two-photon direct laser writing of ultracompact multi-lens objectives *Nat. Photonics* **10** 554–60

[14] Thiele S, Arzenbacher K, Gissibl T, Giessen H and Herkommer A M 2017 3D-printed eagle eye: compound microlens system for foveated imaging *Sci. Adv.* **3** e1602655

[15] Jonušauskas L, Gailevičius D, Rekštytė S, Baldacchini T, Juodkazis S and Malinauskas M 2019 Mesoscale laser 3D printing *Opt. Express* **27** 15205–21

[16] Ristok S, Thiele S, Toulouse A, Herkommer A M and Giessen H 2020 Stitching-free 3D printing of millimeter-sized highly transparent spherical and aspherical optical components *Opt. Mater. Express* **10** 2370–8

[17] Chen X *et al* 2018 High-speed 3D printing of millimeter-size customized aspheric imaging lenses with sub 7 nm surface roughness *Adv. Mater.* **30** 1705683

[18] Shao G, Hai R and Sun C 2020 3D printing customized optical lens in Minutes *Adv. Opt. Mater.* **8** 1901646

[19] Rackson C M *et al* 2021 Object-space optimization of tomographic reconstructions for additive manufacturing *Addit. Manuf.* **48** 102367

[20] Shusteff M *et al* 2017 One-step volumetric additive manufacturing of complex polymer structures *Sci. Adv.* **3** eaao5496

[21] Regehly M *et al* 2020 Xolography for linear volumetric 3D printing *Nature* **588** 620–4

[22] Kelly B E, Bhattacharya I, Heidari H, Shusteff M, Spadaccini C M and Taylor H K 2019 Volumetric additive manufacturing via tomographic reconstruction *Science* **363** 1075–9

[23] Peng S *et al* 2023 Ultra-fast 3D printing of assembly—free complex optics with sub-nanometer surface quality at mesoscale *Int. J. Extreme Manuf.* **5** 035007

[24] Gissibl T, Thiele S, Herkommer A and Giessen H 2016 Sub-micrometre accurate free-form optics by three-dimensional printing on single-mode fibres *Nat. Commun.* **7** 11763

[25] Ge Q *et al* 2020 *Projection Micro Stereolithography based 3D Printing and its Applications* (Bristol: IOP Publishing Ltd.)

[26] Nair S P *et al* 2022 3D printing mesoscale optical components with a low-cost resin printer integrated with a fiber-optic taper *ACS Photonics* **9** 2024–31

[27] Aguirre-Aguirre D, Gonzalez-Utrera D, Villalobos-Mendoza B and Díaz-Uribe R 2023 Fabrication of biconvex spherical and aspherical lenses using 3D printing *Appl. Opt.* **62** C14–20

[28] Gonzalez-Utrera D, Villalobos-Mendoza B, Diaz-Uribe R and Aguirre-Aguirre D 2024 Modeling, fabrication, and metrology of 3D printed Alvarez lenses prototypes *Opt. Express* **32** 3512–27

[29] Hou C, Ren Y, Tan Y, Xin Q and Zang Y 2019 Ultra slim optical zoom system using alvarez freeform lenses *IEEE Photonics J.* **11** 6902110

[30] Ikushima A J, Fujiwara T and Saito K 2000 Silica glass: a material for photonics *J. Appl. Phys.* **88** 1201–13

[31] Inamura C, Stern M, Lizardo D, Houk P and Oxman N 2018 Additive manufacturing of transparent glass structures *3D Print. Addit. Manuf.* **5** 269–83

[32] Wen X *et al* 2021 3D-printed silica with nanoscale resolution *Nat. Mater.* **20** 1506–11

[33] Cooperstein I, Shukrun E, Press O, Kamyshny A and Magdassi S 2018 Additive manufacturing of transparent silica glass from solutions *ACS Appl. Mater. Interfaces* **10** 18879–85

[34] Kotz F *et al* 2017 Three-dimensional printing of transparent fused silica glass *Nature* **544** 337–9

[35] Zhu D, Zhang J, Xu Q and Li Y 2024 3D printing of glass aspheric lens by digital light processing *J. Manuf. Process.* **116** 40–7

[36] Raman R *et al* 2016 High-resolution projection microstereolithography for patterning of neovasculature *Adv. Healthc. Mater.* **5** 610–9

[37] Zhou C and Chen Y 2012 Additive manufacturing based on optimized mask video projection for improved accuracy and resolution *J. Manuf. Process.* **14** 107–18

[38] Pan Y, Zhao X, Zhou C and Chen Y 2012 Smooth surface fabrication in mask projection based stereolithography *J. Manuf. Process.* **14** 460–70

[39] Pan Y and Chen Y 2016 Meniscus process optimization for smooth surface fabrication in Stereolithography *Addit. Manuf.* **12** 321–33

[40] Vaidya N and Solgaard O 2018 3D printed optics with nanometer scale surface roughness *Microsyst. Nanoeng.* **4** 18

[41] Shan Y, Hua J and Mao H 2024 3D printing of optical lenses assisted by precision spin coating *Adv. Funct. Mater.* 2407165

[42] Aničin B A, Babović V M and Davidović D M 1989 Fresnel lenses *Am. J. Phys.* **57** 312–6

[43] Nonimaging fresnel lens design *Nonimaging Fresnel Lenses* **83** (Berlin: Springer) 2001 pp. 77–99

[44] Li D, Sawhney M, Kurtz R, Solomon L and Collette J 2014 Impact of the location of a solar cell in relationship to the focal length of a fresnel lens on power production *Energy Power* **4** 1–6

[45] 1982 Fresnel lenses for ultrasonic inspection *J. Acoust. Soc. Am.* **71** 514

[46] Molerón M, Serra-Garcia M and Daraio C 2014 Acoustic fresnel lenses with extraordinary transmission *Appl. Phys. Lett.* **105** 114109

[47] Klamkin M S 1980 Mathematical modelling: die cutting for a fresnel lens *Math. Modelling* **1** 63–9

[48] Zhang C, Cheng G, Edwards P, Zhou M D, Zheng S and Liu Z 2016 G-fresnel smartphone spectrometer *Lab. Chip.* **16** 246–50

[49] Tan N Y J, Zhang X, Neo D W K, Huang R, Liu K and Senthil Kumar A 2021 A review of recent advances in fabrication of optical Fresnel lenses *J. Manuf. Process.* **71** 113–33

[50] Loaldi D, Quagliotti D, Calaon M, Parenti P, Annoni M and Tosello G 2018 Manufacturing signatures of injection molding and injection compression molding for micro-structured polymer Fresnel lens production *Micromachines (Basel)* **9** 653

[51] Ali M, Alam F, Vahdati N and Butt H 2022 3D-printed holographic fresnel lenses *Adv. Eng. Mater.* **24** 2101641

[52] Ali M, Alam F, Ahmed I, AlQattan B, Yetisen A K and Butt H 2021 3D printing of Fresnel lenses with wavelength selective tinted materials *Addit. Manuf.* **47** 295–303

[53] Ali M, Alam F, Fah Y F, Shiryayev O, Vahdati N and Butt H 2022 4D printed thermochromic Fresnel lenses for sensing applications *Composites* B **230** 109514

[54] Stazi F 2019 Transparent envelope *Advanced Building Envelope Components* (Amsterdam: Elsevier) pp. 1–53

[55] Ali M, Alam F and Butt H 2022 Fabrication of 5D Fresnel lenses via additive manufacturing *ACS Mater. Au.* **2** 602–13

[56] Ali M, Al-Rub R K A and Butt H 2024 Development of high-precision Fresnel lenses for alcohol sensing using vat photopolymerization additive manufacturing *Prog. Addit. Manuf.* **10** 1529–45

[57] Swargiary K, Jolivot R and Mohammed W S 2022 Demonstration of a polymer-based single step waveguide by 3D printing digital light processing technology for isopropanol alcohol-concentration sensor *Photonic Sens.* **12** 10–22

[58] Nagata J, Honma S, Morisawa M and Muto S 2007 Development of polymer optical waveguide-type alcohol sensor *Advanced Materials and Devices for Sensing and Imaging III* (Bellingham, WA: SPIE) p. 682920

[59] Morisawa M, Amemiya Y, Kohzu H, Liang C X and Muto S 2001 Plastic optical fibre sensor for detecting vapour phase alcohol *Meas. Sci. Technol.* **12** 877–81 www.iop.org/Journals/mt

[60] Paixão T, Nunes A S, Bierlich J, Kobelke J and Ferreira M S 2022 Fabry-perot interferometer based on suspended core fiber for detection of gaseous ethanol *Appl. Sci.* **12** 726

[61] Aswathy S H, Narendrakumar U and Manjubala I 2020 Commercial hydrogels for biomedical applications *Heliyon* **6** e03719

[62] Musgrave C S A and Fang F 2019 Contact lens materials: a materials science perspective *Materials* **12** 261

[63] Sato T, Uchida R, Tanigawa H, Uno K and Murakami A 2005 Application of polymer gels containing side-chain phosphate groups to drug-delivery contact lenses *J. Appl. Polym. Sci.* **98** 731–5

[64] Rosa dos Santos J F *et al* 2009 Soft contact lenses functionalized with pendant cyclodextrins for controlled drug delivery *Biomaterials* **30** 1348–55

[65] Kita M, Ogura Y, Honda Y, Hyon S-H, Cha W-I and Ikada Y 1990 Evaluation of polyvinyl alcohol hydrogel as a soft contact lens material *Graefes Arch. Clin. Exp. Ophthalmol.* **228** 533–7

[66] Efron N 2017 *Contact Lens Practice* **100** (Edinburgh: Elsevier Health Sciences)

[67] Glazier A N 1997 *Gas Permeable Elastomer Contact Lens Bonded with Titanium and/or Oxides Thereof* US US-5958194-A

[68] Hales R H 1977 Gas permeable cellulose acetate butyrate (CAB) contact lenses *Ann. Ophthalmol.* **9** 1085–90

[69] Itoh K, Inoue Y, Fujiki H and Yoshino M 1993 *Impression Composition* US US-5367001-A

[70] Efron N and Maldonado-Codina C 2017 7.35 Development of contact lenses from a biomaterial point of view: materials, manufacture, and clinical application *Comprehensive Biomaterials II* **6** (Amsterdam: Elsevier) pp. 686–714

[71] Morrill T J 1998 *Method of Cast Molding Contact Lenses* US US-5843346-A

[72] Mutlu Z, Shams Es-haghi S and Cakmak M 2019 Recent trends in advanced contact lenses *Adv. Healthc. Mater.* **8** 1801390

[73] Han J, Li L and Lee W 2019 Machining of lenticular lens silicon molds with a combination of laser ablation and diamond cutting *Micromachines (Basel)* **10** 250

[74] Chandrinos A and Tzamouranis D-D 2019 Dry eye, contact lenses and preservatives in glaucoma medication mini review *Clin. Ophthalmol. J.* **1** 1003

[75] Hee Keum D *et al* 2020 Wireless smart contact lens for diabetic diagnosis and therapy *Appl. Sci. Eng.* **6** eaba3252

[76] Wang Y, Zhao Q and Du X 2020 Structurally coloured contact lens sensor for point-of-care ophthalmic health monitoring *J. Mater. Chem. B* **8** 3519–26

[77] Alqattan B, Yetisen A K and Butt H 2018 Direct laser writing of nanophotonic structures on contact lenses *ACS Nano* **12** 5130–40

[78] Alam F *et al* 2021 3D printed contact lenses *ACS Biomater. Sci. Eng.* **7** 794–803

[79] Jiang N, Montelongo Y, Butt H and Yetisen A K 2018 Microfluidic contact lenses *Small* **14** 1704363

[80] AlQattan B, Butt H, Sabouri A, Yetisen A K, Ahmed R and Mahmoodi N 2016 Holographic direct pulsed laser writing of two-dimensional nanostructures *RSC Adv.* **6** 111269–75

[81] Neitz J and Neitz M 2011 The genetics of normal and defective color vision *Vision Res.* **51** 633–51

[82] Oli A and Joshi D 2019 Efficacy of red contact lens in improving color vision test performance based on Ishihara, Farnsworth D15, and Martin Lantern Test *Med. J. Armed. Forces. India* **75** 458–63

[83] Seebeck A 1837 Ueber den bei manchen Personen vorkommenden Mangel an Farbensinn *Ann. Phys.* **118** 177–233 (in German)

[84] Swarbrick H A, Nguyen P, Nguyen T and Pham P 2001 The ChromaGen contact lens system: colour vision test results and subjective responses *Ophthalm. Physiol. Opt.* **21** 182–96

[85] Salih A E, Elsherif M, Alam F, Yetisen A K and Butt H 2021 Gold nanocomposite contact lenses for color blindness management *ACS Nano* **15** 4870–80

[86] Alam F, Salih A E, Elsherif M, Yetisen A K and Butt H 2022 3D printed contact lenses for the management of color blindness *Addit. Manuf.* **49** 102464

[87] Hisham M, Salih A E and Butt H 2023 3D printing of multimaterial contact lenses *ACS Biomater. Sci. Eng.* **9** 4381–91

[88] Badawy A R, Hassan M U, Elsherif M, Ahmed Z, Yetisen A K and Butt H 2018 Contact lenses for color blindness *Adv. Healthc. Mater.* **7** 1800152

[89] Stapleton F *et al* 2021 CLEAR—contact lens complications *Cont. Lens Anterior Eye.* **44** 330–67

[90] Baxter S A and Laibson P R 2004 Punctal plugs in the management of dry eyes *Ocul. Surf.* **2** 255–65

[91] Brinton M *et al* 2017 Enhanced tearing by electrical stimulation of the anterior ethmoid nerve *Invest. Ophthalmol. Vis. Sci.* **58** 2341–8

[92] Lee S *et al* 2017 Smart contact lenses with graphene coating for electromagnetic interference shielding and dehydration protection *ACS Nano.* **11** 5318–24

[93] Kusama S, Sato K, Yoshida S and Nishizawa M 2020 Self-moisturizing smart contact lens employing electroosmosis *Adv. Mater. Technol.* **5** 1900889

[94] Aravind M, Chidangil S and George S D 2022 Self-moisturizing contact lens employing capillary flow *Addit. Manuf.* **55** 102842

[95] ter Horst R, Tromp N, de Haan M, Navarro R, Venema L and Pragt J 2017 Directly polished lightweight aluminum mirror *Int. Conf. on Space Optics* (Bellingham, WA: SPIE-International Society of Optical Engineering) p 105660P

[96] Moeggenborg K J, Barros C, Lesiak S, Naguib N and Reggie S 2008 Low-scatter bare aluminum optics via chemical mechanical polishing *Current Developments in Lens Design and Optical Engineering IX* (Bellingham, WA: SPIE) p. 706002

[97] Moeggenborg K, Vincer T, Lesiak S and Salij R 2006 Super-polished aluminum mirrors through the application of chemical mechanical polishing techniques *Current Developments in Lens Design and Optical Engineering VII* (Bellingham, WA: SPIE) p 62880L

[98] Hilpert E, Hartung J, von Lukowicz H, Herffurth T and Heidler N 2019 Design, additive manufacturing, processing, and characterization of metal mirror made of aluminum silicon alloy for space applications *Opt. Eng.* 092613

[99] Zhang K *et al* 2021 Design and fabrication technology of metal mirrors based on additive manufacturing: a review *Appl. Sci.* **11** 10630

[100] Clampin M 2008 The James Webb Space Telescope (JWST) *Adv. Space Res.* **41** 1983–91

[101] Sampath D, Akerstrom A, Barry M, Guregian J, Schwalm M and Ugolini V 2010 The WISE telescope and scanner: design choices and hardware results *An Optical Believe It or Not: Key Lessons Learned II* **7796** 779609

[102] Zhang C and Li Z 2022 A review of lightweight design for space mirror core structure: tradition and future *Machines* **10** 1066

[103] Zhang K *et al* 2021 Design and fabrication technology of metal mirrors based on additive manufacturing: a review *Appl. Sci.* **11** 10630

[104] Atkins C and van de Vorst B 2021 OPTICON A2IM Cookbook: an introduction to additive manufacture for astronomy *Zenedo*

[105] Alzahrani M, Choi S K and Rosen D W 2015 Design of truss-like cellular structures using relative density mapping method *Mater. Des.* **85** 349–60

[106] Hilpert E, Hartung J, Risse S, Eberhardt R and Tünnermann A 2018 Precision manufacturing of a lightweight mirror body made by selective laser melting *Precis. Eng.* **53** 310–7

[107] Hilpert E, Hartung J, von Lukowicz H, Herffurth T and Heidler N 2019 Design, additive manufacturing, processing, and characterization of metal mirror made of aluminum silicon alloy for space applications *Opt. Eng.* **58** 1

[108] Herzog H *et al* 2015 Optical fabrication of lightweighted 3D printed mirrors *Optomechanical Engineering 2015* (Bellingham, WA: SPIE) p. 957308

[109] Mici J, Rothenberg B, Brisson E, Wicks S and Stubbs D M 2015 Optomechanical performance of 3D-printed mirrors with embedded cooling channels and substructures *Optomechanical Engineering 2015* (Bellingham, WA: SPIE) p. 957306

[110] Wang Y, Wu X, Xu L, Ding J, Ma Z and Xie Y 2018 Fabrication of a lightweight Al alloy mirror through 3D printing and replication methods *Appl. Opt.* **57** 8096–101

[111] Guo X, Zhou J, Zhang W, Du Z, Liu C and Liu Y 2017 Self-supporting structure design in additive manufacturing through explicit topology optimization *Comput. Methods. Appl. Mech. Eng.* **323** 27–63

[112] Guo X, Zhang W and Zhong W 2014 Doing topology optimization explicitly and geometrically-a new moving morphable components based framework *J. Appl. Mech. Trans. ASME* **81** 081009

[113] Zhang W, Yang W, Zhou J, Li D and Guo X 2017 Structural topology optimization through explicit boundary evolution *J. Appl. Mech. Trans. ASME* **84** 011011

[114] Yang D, Pan C, Zhou Y and Han Y 2022 Optimized design and additive manufacture of double-sided metal mirror with self-supporting lattice structure *Mater. Des.* **219** 110759

[115] Tian Z, Yang Y, Wang Y, Wu H, Liu W and Wu S 2019 Fabrication and properties of a high porosity h-BN–SiO$_2$ ceramics fabricated by stereolithography-based 3D printing *Mater. Lett.* **236** 144–7

[116] Zhang K *et al* 2019 High solid loading, low viscosity photosensitive Al$_2$O$_3$ slurry for stereolithography based additive manufacturing *Ceram. Int.* **45** 203–8

[117] Xing H *et al* 2018 Preparation and characterization of UV curable Al$_2$O$_3$ suspensions applying for stereolithography 3D printing ceramic microcomponent *Powder Technol.* **338** 153–61

[118] Zhang K, He R, Ding G, Feng C, Song W and Fang D 2020 Digital light processing of 3Y-TZP strengthened ZrO2 ceramics *Mater. Sci. Eng. A* **774** 138768

[119] Zhang K *et al* 2019 Photosensitive ZrO$_2$ suspensions for stereolithography *Ceram. Int.* **45** 12189–95

[120] Wang Z, Huang C, Wang J and Zou B 2019 Development of a novel aqueous hydroxyapatite suspension for stereolithography applied to bone tissue engineering *Ceram. Int.* **45** 3902–9

[121] Ferrage L, Bertrand G, Lenormand P, Grossin D and Ben-Nissan B 2017 *A Review of the Additive Manufacturing (3DP) of Bioceramics: Alumina, Zirconia (PSZ) and Hydroxyapatite* (Cham: Springer International Publishing)

[122] Zhang Y, Zhang J, Han J, He X and Yao W 2004 Large-scale fabrication of lightweight Si/SiC ceramic composite optical mirror *Mater. Lett.* **58** 1204–8

[123] Liu G, Zhang X, Yang J and Qiao G 2019 Recent advances in joining of SiC-based materials (monolithic SiC and SiCf/SiC composites): joining processes, joint strength, and interfacial behavior *J. Adv. Ceram.* **8** 19–38

[124] Li S, Zhang Y, Han J and Zhou Y 2013 Fabrication and characterization of SiC whisker reinforced reaction bonded SiC composite *Ceram. Int.* **39** 449–55

[125] Yang W S *et al* 2012 Fabrication of short carbon fibre reinforced SiC multilayer composites by tape casting *Ceram. Int.* **38** 1011–8

[126] Zhang J, Jiang D, Lin Q, Chen Z and Huang Z 2013 Gelcasting and pressureless sintering of silicon carbide ceramics using Al$_2$O$_3$–Y$_2$O$_3$ as the sintering additives *J. Eur. Ceram. Soc.* **33** 1695–9

[127] Franks G V, Tallon C, Studart A R, Sesso M L and Leo S 2017 Colloidal processing: enabling complex shaped ceramics with unique multiscale structures *J. Am. Ceram. Soc.* **100** 458–90

[128] Ding G, He R, Zhang K, Zhou N and Xu H 2020 Stereolithography 3D printing of SiC ceramic with potential for lightweight optical mirror *Ceram. Int.* **46** 18785–90

[129] Zhang C, Anzalone N C, Faria R P and Pearce J M 2013 Open-source 3D-printable optics equipment *PLoS One* **8** e59840

[130] da Cruz Junior L B and Bachmann L 2021 Manufacture and characterization of a 3D-printed integrating sphere *Instrum Sci. Technol.* **49** 276–87

[131] Salazar-Serrano L J, P. Torres J and Valencia A 2017 A 3D printed toolbox for opto-mechanical components *PLoS One* **12** e0169832

[132] Tomes J J and Finlayson C E 2016 Low cost 3D-printing used in an undergraduate project: an integrating sphere for measurement of photoluminescence quantum yield *Eur. J. Phys.* **37** 055501

[133] Prahl S A, van Gemert M J C and Welch A J 1993 Determining the optical properties of turbid media by using the adding–doubling method *Appl. Opt.* **32** 559–68

[134] Salazar-Serrano L J, Jiménez G and Torres J P 2018 How to automate a kinematic mount using a 3D printed arduino-based system *Inventions* **3** 39

[135] Gunderson J E C, Mitchell D W, Bullis R G, Steward J Q and Gunderson W A 2020 Design and implementation of three-dimensional printable optomechanical components *J. Chem. Educ.* **97** 3673–82

[136] Diederich B *et al* 2020 A versatile and customizable low-cost 3D-printed open standard for microscopic imaging *Nat. Commun.* **11** 5979

[137] Delmans M and Haseloff J 2018 μCube: a framework for 3D printable optomechanics *J. Open Hardware* **2** 2

[138] Winters B J and Shepler D 2018 3D printable optomechanical cage system with enclosure *HardwareX* **3** 62–81

[139] Stewart C and Giannini J 2016 Inexpensive, open source epifluorescence microscopes *J. Chem. Educ.* **93** 1310–5

[140] Chagas A M, Prieto-Godino L L, Arrenberg A B and Baden T 2017 The €100 lab: a 3D-printable open-source platform for fluorescence microscopy, optogenetics, and accurate temperature control during behaviour of zebrafish, Drosophila, and *Caenorhabditis elegans* *PLoS Biol.* **15**

[141] Grasse E K, Torcasio M H and Smith A W 2016 Teaching UV–Vis spectroscopy with a 3D-printable smartphone spectrophotometer *J. Chem. Educ.* **93** 146–51

[142] Porter L A, Chapman C A and Alaniz J A 2017 Simple and inexpensive 3D printed filter fluorometer designs: user-friendly instrument models for laboratory learning and outreach activities *J. Chem. Educ.* **94** 105–11

[143] Mantia M and Bixby T 2022 Optical measurements on a budget: a 3D-printed ellipsometer *Am. J. Phys.* **90** 445–51

[144] Bernard P and Mendez J D 2020 Low-cost 3D-printed polarimeter *J. Chem. Educ.* **97** 1162–6

[145] James S W and Tatam R P 2003 Optical fibre long-period grating sensors: characteristics and application *Meas. Sci. Technol.* **14** 49–61

[146] Erdogan T 1997 Fiber grating spectra *J. Lightwave Technol.* **15** 1277–94

[147] Vengsarkar A M, Bergano N S, Davidson C R, Pedrazzani J R, Judkins J B and Lemaire P J 1996 Long-period fiber-grating-based gain equalizers *Opt. Lett.* **21** 336–8

[148] Gu X J 1998 Wavelength-division multiplexing isolation fiber filter and light source using cascaded long-period fiber gratings *Opt. Lett.* **23** 509–10

[149] Ramachandran S, Wang Z and Yan M 2002 Bandwidth control of long-period grating-based mode converters in few-mode fibers *Opt. Lett.* **27** 698–700

[150] Bhafia V and Vengsarkar A M 1996 Optical fiber long-period grating sensors *Opt. Lett.* **21** 692–4

[151] Bae J K, Kim S H, Kim J H, Lee S B, Jeong J M and Bae J H 2003 Spectral shape tunable band-rejection filter using long period fiber gratings with divided coil heater *IEEE Photonics Technol. Lett.* **15** 407–9

[152] Coelho J M P, Silva C, Nespereira M, Abreu M and Rebordão J 2015 Writing of long period fiber gratings using CO_2 laser radiation *Advances in Optical Fiber Technology: Fundamental Optical Phenomena and Applications* (London: IntechOpen) 287–314

[153] Vengsarkar A M, Lemaire P J, Judkins J B, Bhatia V, Erdogan T and Sipe J E 1996 Long period fiber gratings as band-rejection filters *J. Lightwave Technol.* **14** 58–65

[154] Li B, Jiang L, Wang S, Tsai H L and Xiao H 2011 Femtosecond laser fabrication of long period fiber gratings and applications in refractive index sensing *Opt. Laser Technol.* **43** 1420–3

[155] Von Bibra M L, Roberts A and Canning J 2001 Fabrication of long-period fiber gratings by use of focused ion-beam irradiation *Opt. Lett.* **26** 765–7

[156] Colaço C, Caldas P, Del Villar I, Chibante R and Rego G 2016 Arc-induced long-period fiber gratings in the dispersion turning points *J. Lightwave Technol.* **34** 4584–90

[157] Oliveira R, Sousa L M, Rocha A M, Nogueira R and Bilro L 2021 UV inscription and pressure induced long-period gratings through 3D printed amplitude masks *Sensors* **21** 1977

[158] Tachibana M, Laming R I, Morkel P R and Payne D N 1991 Erbium-doped fiber amplifier with flattened gain spectrum *IEEE Photonics Technol. Lett.* **3** 118–20

[159] Iezzi V L, Boisvert J-S, Loranger S and Kashyap R 2016 3D printed long period gratings for optical fibers *Opt. Lett.* **41** 1865–8

[160] Khun-In R, Usuda Y, Jiraraksopakun Y, Bhatranand A and Yokoi H 2020 Resin made long-period fiber grating structure for tunable optical filter inside single-mode fiber *Key Engineering Materials* (Bäch: Trans Tech Publications Ltd) pp. 259–63

[161] Lee J, Kim Y and Lee J H 2020 A 3D printed, temperature sensor based on mechanically-induced long period fibre gratings *J. Mod. Opt.* **67** 469–74

[162] Oliveira R, Nogueira R and Bilro L 2021 3D printed long period gratings and their applications as high sensitivity shear-strain and torsion sensors *Opt. Express* **29** 17795–814

[163] Valente N F, Bilro L and Oliveira R 2021 3D printing of long period gratings for curvature applications *EPJ Web. Conf.* **255** 12001

[164] Vaziri M and Chen C-L 1992 Etched fibers as strain gauges *J. Lightwave Technol.* **10** 836–41

[165] Lin C Y and Wang L A 1999 Loss-tunable long period fibre grating made from etched corrugation structure *Electron. Lett.* **35** 1872–3

[166] Chiang C-C 2010 Novel fabrication method of corrugated long-period fiber gratings by thick SU-8 photoresist and wet-etching technique *J. Micro/Nanolithogr. MEMS MOEMS* **9** 033007

[167] Valente N F, Figueira F, Bilro L and Oliveira R 2024 Strain, displacement, and temperature opportunities of a periodically etched optical fiber *IEEE Sens. J.* **24** 2871–9

IOP Publishing

Additive Manufacturing in Optics and Photonics
Fabrication and applications
Ricardo Oliveira and Nuno Valente

Chapter 3

Three-dimensional printing of optical waveguides

3.1 Optical waveguides

Optical waveguides are fundamental structures in photonic systems, enabling the propagation of light radiation along a specific path with minimal insertion losses. These structures guide electromagnetic waves using total internal reflection (TIR) within a high-refractive-index core surrounded by a lower-refractive-index cladding. Today, waveguides serve as the backbone of our optical communication infrastructure, as well as in integrated photonics and sensing.

Waveguides come in various types, including planar, slab, and fibre-based structures. Planar waveguides, commonly used in photonic integrated circuits (PICs), guide light in a two-dimensional plane, while optical fibres guide light in three dimensions along their length. Additionally, waveguides can be designed with periodic structures to confine light along the direction of propagation. Examples of these advanced waveguides include photonic crystal waveguides and grating-based waveguides, which enable light confinement through Bragg reflection and bandgap engineering.

Optical waveguides involve complex and delicate fabrication methodologies that enable the manipulation of mode propagation with minimal insertion loss. This is generally achieved through precision technologies capable of fabricating designs at the nanoscale.

Additive manufacturing (AM) has shown unparalleled design flexibility and rapid fabrication capabilities. Unlike traditional fibre-drawing methods and lithographic techniques, AM enables the creation of micro- to nanoscale free-form optical structures, marking a new era in waveguide development. To achieve this, transparent polymer and silica materials have been utilised in various manufacturing technologies, including vat polymerisation methods such as stereolithography (SLA) and two-photon polymerisation (TPP). While the former facilitates the production of large-scale waveguides with micrometre resolution, typically functioning in the multimode (MM) regime, the latter is better suited for fabricating small, intricate sub-micrometre features operating in the single-mode (SM) regime.

The fabrication of optical waveguides through AM can create stand-alone optical elements, such as splitters, combiners, and mode converters, or can be developed on various substrates or optical components, including PICs and optical fibres. Additionally, their use in photonic bond wires is also very appealing, allowing, for instance, PIC-PIC interconnects and PIC-fibre interconnects, with minimal alignment challenges. Nonetheless, special optical fibre designs that were once theoretically idealised but impossible to manufacture through traditional methods can now be produced through preform AM fabrication followed by standard fibre-drawing processes. With AM techniques, optical waveguides are now capable of bringing new technological advances to the optics and photonics world.

3.2 Types of waveguides

The most widely used waveguides, particularly those relying on optical waveguiding, operate according to Snell's law, which governs the behaviour of light as it propagates through materials with varying refractive indices. This law states that when light travels from a medium with a higher refractive index to one with a lower refractive index, it undergoes refraction. If the incident angle exceeds a critical threshold, known as the critical angle (θ_c), the light is entirely reflected at the interface rather than refracted, resulting in TIR. This principle forms the basis of optical confinement in waveguides.

The simplest form of optical waveguide is the planar waveguide, which consists of a slab geometry. In its most basic configuration, the core layer, composed of high-refractive-index material, is deposited on a substrate, with the surrounding medium being either air or another dielectric material. A more common variation of the slab waveguide consists of a core embedded between two materials: a lower-refractive-index cladding and the substrate. In this case, the refractive index of the core (n_2) must be higher than those of both the substrate (n_3) and the cladding (n_1), ensuring optical confinement through TIR.

To maintain guided modes, the thickness of both the cladding and the substrate must be sufficient to prevent significant penetration of the evanescent field outside the core. In a symmetric slab waveguide, where the refractive index of the cladding is equal to that of the substrate ($n_1 = n_3$), the refractive index profile becomes symmetric, thereby simplifying the analysis of wave propagation. In this case, the minimum angle required for TIR can be determined using equation (3.1):

$$\theta_c = \sin^{-1}\left(\frac{n_1}{n_2}\right) \tag{3.1}$$

where n_1 is the refractive index of the cladding/substrate, and n_2 is the refractive index of the core.

Beyond planar waveguides, nonplanar optical waveguides offer light confinement in two transverse directions (x and y). Here, the core is enveloped by cladding on all transverse sides. Examples of these waveguides include channel waveguides and optical fibres. These waveguides enable more complex manipulations of light, thereby opening the door to advanced photonic devices.

3.3 Three-dimensional printing in photonic integrated circuits

PICs are becoming more popular, as they facilitate high-speed communication, enabling data transmission over long distances with lower power consumption than existing electrical solutions [1, 2]. In a PIC, various components and systems are integrated into densely populated substrates, resulting in especially compact components. This type of device has already been reported in several applications, such as sensing and communications [3–5]. The PIC industry utilises semiconductor fabrication technologies, initially developed for electronic integrated circuits, reducing the manufacturing costs of photonic device components [6]. Unfortunately, not all processes can take advantage of this opportunity. For instance, the packaging process requires precise alignment of the PIC with the optical fibre/fibre bundle, which must comply with strict conditions to ensure proper matching of the optical modes [7, 8]. These alignment requirements demand micron or submicron-level tolerances, making the related processes both complex and costly, and rendering them incompatible with conventional electronic packaging techniques, given that their tolerances are less stringent [9, 10]. As a result, the search for alternative packaging methods suitable for such small scales has spurred innovation, leading to the adoption of new packaging technologies. 3D printing technologies have become a promising solution, especially two-photon lithography, which is capable of printing at resolutions below the diffraction limit [11], making it an excellent option for the photonic packaging of PICs [12], thus potentially addressing the packaging challenges.

3.3.1 Photonic wire bonding

Currently, photonic integration is one of the most important technologies in high-speed telecommunications and quantum information processing. As a result, several technologies have been developed in the field of integrated photonics. In the area of photonic chip assembly, various coupling elements have been developed, including mode-field adaptors [13], microlenses, and mirrors [14]. Despite the usefulness of these components, they pose significant limitations, including the complexity of aligning these components with arrays of chips. This raises the cost of these processes, undermining some of the benefits of using PICs. As a result, various approaches for photonic integration have been investigated, one of which is monolithic integration [15], i.e. the integration of all elements on the same substrate. Although these types of PICs reduce the number of connections and thereby reduce costs and complexity, they remain constrained by the optical properties of the substrate material. The limitations of the previously described technologies have led to the exploration of new solutions that build upon existing methods, such as AM. One solution currently being investigated for the production of PICs is photonic wire bonding (PWB). This concept was first explored in 2012 [16] and involves the TPP direct laser writing (TPP-DLW) fabrication of a polymer link between two optical waveguides through a polymeric interlink. A schematic example of PWBs acting as fibre-to-chip and chip-to-fibre interconnects in a multi-chip system is shown in the schematic of figure 3.1 [16].

As shown in figure 3.1, the PWB serves as a photonic interconnect between chips and fibres. The fabrication of a PWB requires various steps. The first involves fixing the chips and fibres to a common sub-mount using pick-and-place machinery. Then, the interconnect regions are immersed in a negative resist ($n = 1.57$ @ 1550 nm), and the positions of the facet waveguide structures are detected within the resin. Later, the TPP fabrication process is used to create PWB interconnects with the desired patterns and dimensions. A washing process is then carried out to remove the unpolymerised resist. Finally, the structure is immersed in an index-matching liquid with $n = 1.30$ (@ 1550 nm), allowing it to emulate the low-index cladding material

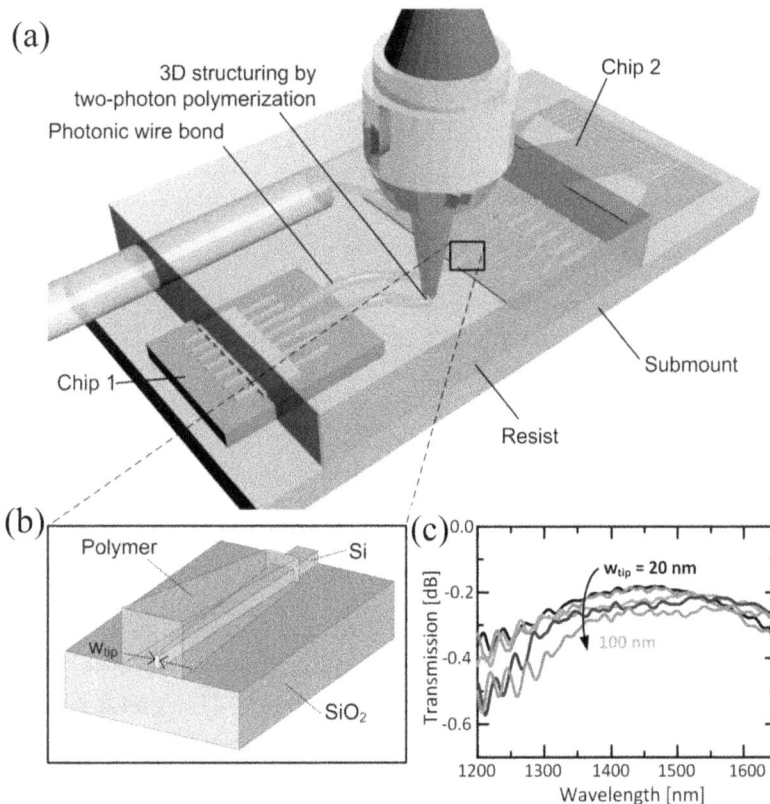

Figure 3.1. (a) Two PICs and an optical fibre are mounted on a sub-mount and immersed in a negative resist. A femtosecond laser beam is focused in 3D space, creating a photonic wire bond (PWB) between the different waveguides. (b) Transition between a silicon-on-insulator nanowire and a PWB interconnect. The silicon waveguide is 500 nm wide and 220 nm high and is tapered down along a 32 μm length to a tip width, w_{tip}, varying between 20 and 100 nm. (c) Simulated transmission loss spectra for tip widths of different sizes. Reprinted with permission from [16]. © 2012 Optical Society of America.

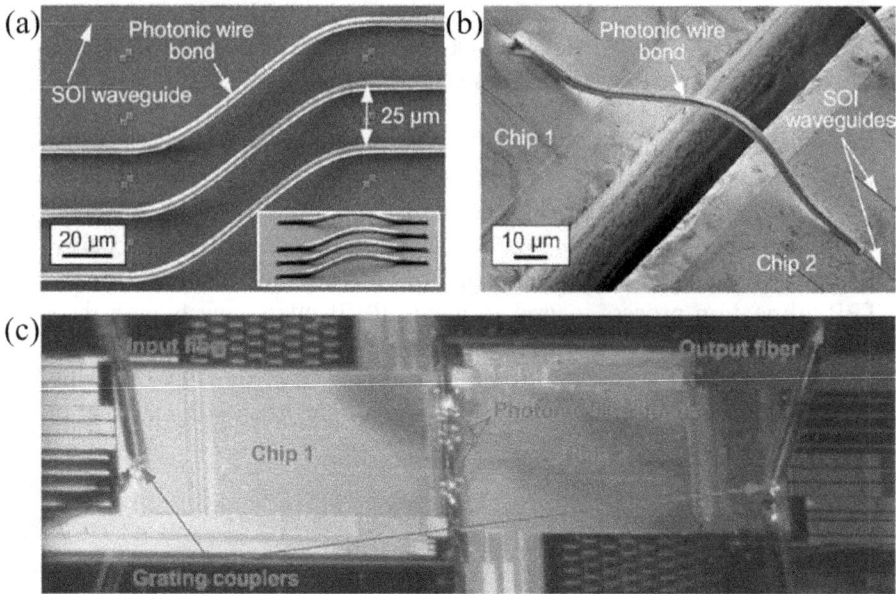

Figure 3.2. (a) Photonic wire bonds between two SOI waveguides on the same chip. (b) PWB between two different chips. (c) Experimental setup used to characterise the PWB between two SOI chips. The output and input of the setup are connected using two SM fibres. Reprinted with permission from [16]. © 2012 Optical Society of America.

[16]. The results of this process are shown in figure 3.2, where the PWB between two silicon-on-insulator (SOI) waveguides is clearly visible.

The PWB depicted in figure 3.2(b) and presented in [16] illustrates the technology's capability to manage tip displacements, as the connected tips were spaced 100 μm apart, featuring a lateral shift of 25 μm horizontally and 12 μm vertically. The bridges demonstrated solid mechanical and chemical characteristics, as the photonic wire bonds produced had a diameter of just 2 μm, and the arcs of these waveguides remained intact over distances greater than 100 μm. The wire bonds also revealed excellent adhesion to the SOI waveguides and great resistance to humidity and oxygen.

Concerning the PWB transmission losses, they were estimated to be 3 ± 1 dB for the wavelength range of 1240–1580 nm [16]. These results revealed that PWBs offer higher efficiency than the use of fibre–chip couplers [17] and planar coupling schemes based on electron-beam lithography [18]. Regarding the use of PICs in systems with multi-terabit s^{-1} data streams, it was revealed that a PIC with PWB achieved an aggregate transmission rate of 5.25 Tbit s^{-1} in a wavelength division multiplexing (WDM) system, being one of the highest rates obtained for an SOI nanowire.

PWB proves to be an efficient path towards interconnects with spatial densities in the Pbit s^{-1} mm range [16], allowing easy integration of different photonic device technologies and their implementation in large-scale photonic integration.

3.3.2 Photonic wire bonding in hybrid photonic multi-chip modules

Due to its indirect bandgap, silicon cannot be used for the fabrication of efficient optical sources. On the other hand, III–V materials, such as GaAs (gallium arsenide), InP (indium phosphide), InGaAsP, AlGaAs, etc. have direct bandgaps, making them very efficient at emitting light. Thus, the integration of III–V-based sources (laser diodes or light-emitting diodes) into an SOI platform is seen as the most attractive solution [19–21]. This integration can be performed using two approaches: heterogeneous integration [22] and hybrid integration [14]. In heterogeneous integration, the III–V dies are bonded to silicon photonics (SiP) wafers, and the light generated in the III–V die is evanescently coupled to the SiP [23]. Although this technique is extremely useful for mass production, since it does not require high-precision positioning of the dies during the bonding process, it does not allow the light sources to be tested before they are integrated into the SiP wafers. These limitations demonstrate that heterogeneous integration is primarily helpful in cases where a large number of light sources need to be integrated into a single chip and where the integration process is well established. Hybrid integration involves optically connecting various elements, such as III–V lasers, photodiodes, or other components, to SiP circuits [24]. In hybrid integration, it is typically necessary to align the components at the submicron scale, making this process complex and time-consuming. In this process of alignment, it is sometimes necessary to use external components such as lenses or mirrors to adapt the mode-field diameters of the different elements, leading to the assembly of complex and bulky setups. The limitations of both the existing methods of heterogeneous integration and hybrid integration led to the integration of AM into the process of photonic integration. PWB has already demonstrated its capability to connect photonic elements with relatively low losses; therefore, Billah *et al* [25] proposed the use of PWBs to create a hybrid photonic multi-chip module. They proposed the assembly of passive SiP circuits with InP light sources using PWB. A representative schematic of the proposed hybrid system and associated scanning electron microscope (SEM) images of the assembly can be observed in figure 3.3.

To demonstrate the scalability of the technology, the authors also manufactured an array of PWBs connecting several distributed feedback (DFB) lasers to SiP waveguides [25]. To determine whether the PWB was a viable solution for connecting the laser sources to the chips, the insertion losses between an InP laser and SiP waveguides were measured. These connections presented low coupling losses, $\approx$0.4 dB, showing the high potential of the technology for photonic integration compared to its competitors, which presented insertion losses as high as 2.3 dB [13].

The proof of concept for the use of PWBs in a hybrid photonic multi-chip module described in [25] demonstrated efficient coupling between components in complex photonic setups. However, since the proposed study cases were observed in controlled environments, there was still a need to prove the reliability of PWBs in real-world scenarios. For this reason, two years later, the same group reported the full characterisation of two complex optical communication photonic setups relying

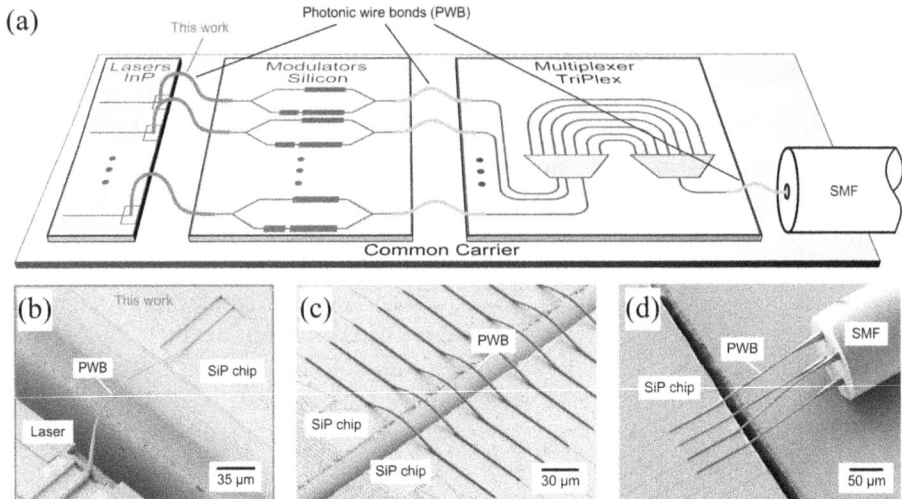

Figure 3.3. (a) System concept for WDM communications that uses PWB to connect different photonic chips in a hybrid photonic multi-chip module. (b) Scanning electron microscope (SEM) image of a laser-to-chip connection. (c) SEM image of a chip-to-chip connection. (d) SEM image of a fibre-to-chip connection. Parts (a)–(c) reprinted with permission from [25] © 2018 Optical Society of America. Part (d) © [2015] IEEE. Reprinted, with permission, from [26].

on photonic wire bonds to connect arrays of SiP modulators to InP lasers and single-mode fibres (SMFs) [27].

The first hybrid multi-chip transmitter engine realised with photonic wire bonds consisted of an array of eight InP lasers connected to eight different SiP modulators, which were connected to an array of eight SMF fibres. The modulators of the system were driven by an arbitrary waveform generator (AWG), providing two types of signals: two-level on-off-keying (OOK) or four-level pulse amplitude modulation (PAM-4). When all channels were used to carry OOK signals, an aggregate line rate of 320 Gbit s^{-1} was obtained for a transmission distance of 2 km. When PAM-4 signals were carried, researchers obtained an aggregate line rate of 448 Gbit s^{-1} over a distance of 10 km, leading to compact high-speed 400 Gbit s^{-1} modules as specified in various standards [27].

The second hybrid multi-chip transmitter engine that was developed consisted of an array of four InP lasers that fed the SiP chip, composed of four silicon–organic hybrid IQ modulators. The SiP chip was connected to four SMF fibres, each with a length of 75 km. All these elements were connected by PWB. The proposed system presented promising results, such as low energy consumption and an aggregate line rate of 784 Gbit s^{-1} over a distance of 75 km, which was one of the highest presented in the research literature.

To verify the reproducibility of the PWB, the authors fabricated a total of 100 densely spaced photonic wire bond bridges with a pitch of 25 μm on a single chip. The results showed average losses of 0.7 dB and a standard deviation of 0.2 dB. Additionally, to verify the reliability of the photonic wire bonds, they were subjected to multiple temperature cycles ranging from −40 °C to 85 °C, as well as exposure to humidity, without exhibiting any signs of degradation. Finally, the researchers tested the PWB using continuous laser irradiation at 1550 nm and confirmed that the photonic wire bonds were only destroyed by nonlinear absorption at a power of ≈19 dBm. Overall, the results confirmed the reliability of the structures under relevant environmental conditions and at optical power levels realistically achieved in SiP assemblies [27].

3.3.3 Waveguide crossings

The advancement of PIC technology has increasingly complicated these components, enabling the integration of thousands of devices onto a single chip [28]. As the complexity increases, nonplanar topologies, along with numerous waveguide crossings (WGXs), have become a reality. As WGXs proliferate, research focused on enhancing the performance of these components has similarly escalated. Various intriguing results related to insertion losses and crosstalk have been observed for different WGX arrays [29–31]. Despite these favourable outcomes, crosstalk poses significant challenges for PICs with numerous WGXs, potentially resulting in errors that approach the limits of hard-decision forward error correction. To overcome the signal degradation caused by crosstalk in PICs featuring hundreds or thousands of WGXs, one solution that has been explored involves the stacking of multiple waveguide layers constructed from silicon [32], silicon nitride [33], or a combination of these [34, 35]. The implementation of this solution for a two-layer structure offered decent performance [36], while for a three-layer structure, the performance was greatly increased, and the crosstalk diminished significantly [34, 35]. Despite the success of this solution, the use of multiple layers substantially increases process complexity, and most foundries in the market still cannot implement this solution. To overcome this challenge, one potential solution is the use of 3D-printed free-form polymer structures to realise nonplanar circuit typologies, also designated as optical waveguide overpasses (WOPs) [37]. These waveguides are fabricated *in situ* by TPP-DLW, as previously described for the fabrication of photonic wire bonds.

The manufacture of a WOP involves the fabrication of a PIC using standard commercial foundry processes, including selective removal of the cladding region to access the interconnecting tapers needed for WOP fabrication. Then, a negative-tone photoresist is placed on top of the tapered exposed etched regions, and the fabrication of the WOP is followed by direct laser writing using a TPP 3D printer. An example of the process detailed in [37] for a WOP demonstrator in a SiP circuit is shown in figure 3.4(a), while SEM images of the fabricated structures after the unpolymerised resist was washed with developer are shown in figure 3.4(b). Figures 3.4(c)–(e) show an overview and close-ups of different regions of the WOP.

The WOP described in [37] and shown in the SEM images of figure 3.4 was designed with a specific distance, *w*, between the SiP waveguides. This distance can be adjusted depending on the number of planar waveguides that need to be bridged, which can range from several tens to hundreds, depending on the complexity of the

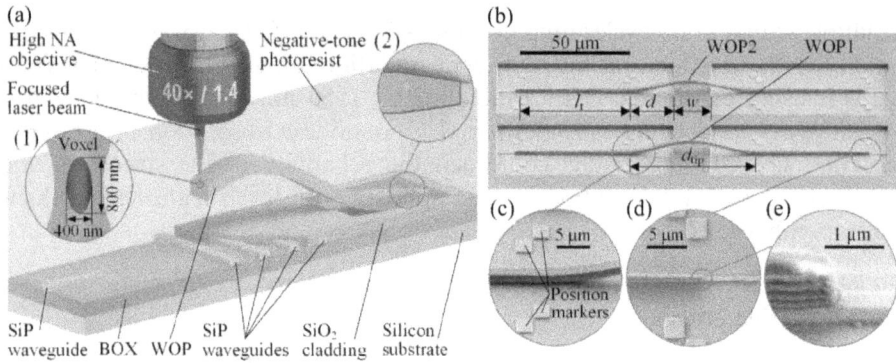

Figure 3.4. (a) Writing a WOP in a liquid negative-tone photoresist. (b) SEM image of the WOP. Panels (c), (d), and (e) are close-ups of the different zones of the WOP. The position markers shown in the close-ups indicate the zone of the taper that needs to be interconnected. Reprinted with permission from [37] © 2019 Optical Society of America.

PIC. However, it is worth mentioning that larger distances are more suitable, as they reduce the insertion losses associated with the curvature radius required by the WOP over shorter distances. In figures 3.4(b)–(d), it is possible to see position markers intentionally included during the complementary metal–oxide–semiconductor (CMOS) patterning that takes place during PIC manufacture, which are used for easy automated detection of the SiP waveguides that need to be interconnected through the TPP-DLW. Another important detail during the fabrication of this WOP is the tapered regions of the SiP and WOP regions shown in the inset of figure 3.4(a), which were optimised for efficient mode coupling between each other. After the printing and cleaning processes are complete, the WOP is cladded with a polymer with a lower refractive index than that of the TPP-fabricated WOP, i.e. 1.53 vs. 1.36 @ 1550 nm, respectively.

According to the study described in [37], as the number of port counts, n, increases in a switch-and-select (SAS) circuit, the number of WGXs increases according to n^4, whereas the number of WOPs only increases in proportion to n^2, making the use of WOPs highly attractive. To demonstrate the viability of integrating 3D-printed free-form WOPs, authors used a 4×4 SAS circuit realised on a SiP platform. The SAS circuit consisted of four 1×4 switches at the input and four 4×1 switches at the output. The device had a total of 24 Mach–Zehnder interferometers (MZIs) and 48 phase shifters (i.e. each MZI had a pair of phase shifters). To evaluate the interconnection, 16 phase shifters were required for simultaneous operation, while the remaining ones were idle. In the described chip, 16 optical paths connected the input to the output switch. To evaluate the performance of the SAS PIC, the transmission spectra of all 16 optical paths were measured with and without WOPs. To perform the measurement, the authors used surface coupling to optical fibres using grating couplers (GCs). The setup used for the characterisation is displayed in figure 3.5(a), while photomicrographs of the PIC and the transmission power results can be seen in figures 3.5(b)–(c) and (d), respectively.

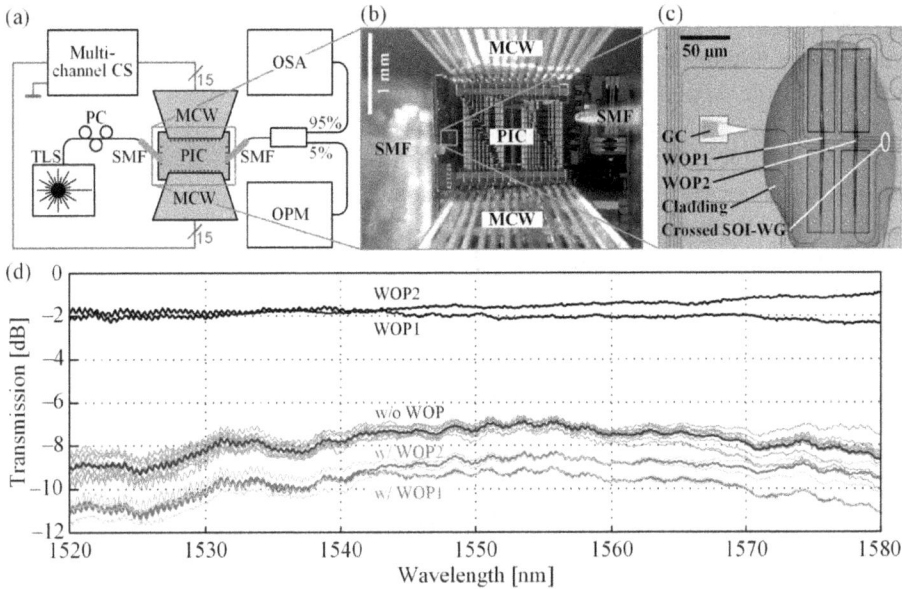

Figure 3.5. (a) Experimental setup used to demonstrate a 4 × 4 SAS circuit with WOPs used to test all 16 possible optical paths that connect the various input and output ports of a 4 × 4 SAS PIC. A representative schematic of the setup used to measure the transmission spectra of the 16 optical paths. (b) Photomicrograph of the SAS PIC showing multi-contact probe wedges (MCWs) and the PIC. (c) Magnified region of the photomicrograph, showing two WOPs bridging three and four SiP strip waveguides. (d) Transmission spectra of the 12 optical paths in the switch with and without WOPs. Reprinted with permission from [37] © 2019 Optical Society of America.

To evaluate the loss performance of the SAS PIC, the fibre coupling losses were discounted. Measurements of the different optical paths revealed that the average losses were ≈7 dB when the WOP was not present in the system. When the WOPs were inserted into the system, the losses increased, namely by 1.6 and 1.9 dB. These high losses were attributed to the non-optimal design of the on-chip coupling structures for the WOP, and also to the small width, $w = 17$ μm, and small distance, d, between the tips of the tapered on-chip SiP waveguides and the edge of the oxide opening (i.e. $d = 20$ μm), leading to a smaller bend radius and, as a result, increased losses. Finally, the authors also measured the crosstalk from a WOP to one of the SiP waveguides underneath it. The results showed low crosstalk, with values of less than −75 dB, which were lower than the values reported in the literature for SiN-based multi-layer circuits. Overall, the authors demonstrated that WOPs are a possible candidate to bridge a series of parallel waveguides, allowing for the efficient replacement of a multitude of WGXs.

3.3.4 PWB between multicore fibre and photonic integrated circuits

Our optical communication systems strongly rely on the use of technologies such as wavelength, polarisation, and space division multiplexing (SDM), among others. In the field of SDM, multicore optical fibres (MCFs) have emerged as a potential solution to address the bottleneck in communication traffic. In an MCF, multiple parallel cores within the same cladding region run along the same fibre cable. This enables multiple optical signals to travel through different cores simultaneously, allowing the fibre to support transmission at high data rates [38].

For MCF deployment, fibres need to be connected to highly integrated photonic transmitter and receiver circuitry. However, since the cores in an MCF are densely packed in the cladding region, it becomes challenging to couple light in and out of them. Fan-in and fan-out devices [39, 40] are possible options, but the idea of not depending on an external device for the MCF-PIC coupling seems much more convenient in terms of cost and volume integration. Although one might easily consider using facet coupling, this would necessitate multiple waveguide layers with layer thicknesses spaced apart by tens of micrometres, making it inappropriate. One possible solution could make use of an out-of-plane connection between the MCF and an SOI PIC based on GCs [41]. However, this approach has considerable drawbacks, including the complex relative alignment of the MCF tip to the PIC across six degrees of freedom, which restricts its application for large-scale production. Furthermore, the use of GCs reduces the transmission bandwidth and prevents the use of dense WDM techniques. Thus, other strategies were sought. One of these consisted of the use of PWB between each of the cores found at the facet of the MCF and the waveguides of a PIC [26]. This study demonstrated the coupling of a four-core MCF to an SOI PIC using the TPP-DLW technique. To achieve this, the PWBs were up-tapered to match the mode-field diameter of the MCF cores and the SOI waveguides of the PIC. For the PWB manufacture, a commercial TPP printer was used. For this purpose, the collimated beam of a femtosecond laser was focused through a 100x lens onto photoresist covering the region intended for the PWB. The voxel size reached values of $\approx$100 nm, and the PWB was created by moving the focal spot of the writing beam through the volume of the photoresist. The final result of the printing process is shown in figure 3.6 [26].

The SEM image presented in figure 3.6 confirms the successful fabrication of photonic wire bonds between an MCF and a SiP waveguide chip [26]. The authors

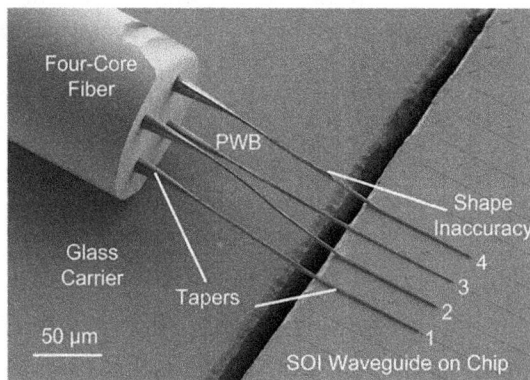

Figure 3.6. Photonic wire bonds made between the four cores of the MCF and the different waveguides of the SiP chip. © [2015] IEEE. Reprinted, with permission, from [26].

estimated theoretical losses of 1.3 dB for the waveguide, which included curved paths and tapering regions. These calculations assumed refractive indices of 1.53 and 1.34 for the PWB and the cladded region, respectively. Of the total loss, 0.8 dB was attributed to the transition between the PWB and the SOI waveguide, while the remainder was attributed to the interface with the MCF. In the experiments, the lowest measured value was 1.7 dB. Although this is higher than the theoretical predictions, likely due to scattering losses from imperfect PWBs, this result remains a noteworthy accomplishment.

3.3.5 Lenses for chip-to-fibre and fibre-to-fibre coupling

One of the biggest challenges when working with PICs is the coupling losses at the optical interface between SMFs and PICs [42]. So far, high-index contrast grating structures at the chip surface [43] have been widely implemented in PIC technology. Despite this, several PIC technologies require coupling to waveguide facets at the chip edge. This can be achieved by adding lenses at the tips of the fibres, which can be designed to match the mode-field diameters of the structures they are intended to couple with. The difficulty associated with this approach lies in the manufacturing process of fibre lenses, which typically involves grinding, etching, and melting [44, 45]. These processes are not easy to handle, making the mass production of such fibres hard to achieve. One possible solution could involve the use of microlenses [46]. However, this solution faces a challenge: the lenses must be assembled to align within a tolerance of three to six degrees of freedom. This problem can be solved by 3D printing [47]. Proofs of concept have been demonstrated for the light coupling between a fibre and a PIC and from one fibre to another [48]. The process consists of fabricating microlenses on the tips of optical fibres. To achieve this, the lenses need to be optimised for the mode-field diameter of the element to be coupled; i.e. for light coupling between standard SMFs, a 10.2 μm diameter should be considered. However, when considering the coupling between a laser diode assembled in a PIC and an SMF, the lens needs to be optimised for the smaller mode-field diameter, i.e. the PIC waveguide. In reference [48], the lenses used for coupling between a fibre and a PIC were designed to have an efficiency of 98% for a mode-field diameter of 2 μm. The position of the lens surface z was described by a rotationally symmetric polynomial surface of the form $z(r) = a_0 + a_2 r^2 + a_4 r^4 + a_6 r^6$, where r is the distance from the optical axis and a_0, a_2, a_4, and a_6 are the numerically optimised coefficients that achieve maximum efficiency. The microlenses were 3D printed using a commercial TPP printer in a negative photoresist, with a writing speed of 10 mm s^{-1} and hatching and slicing distances of 100 nm. Schematics of the couplings between different optical elements can be seen in figures 3.7(a) and (b), while SEM images of the lenses produced in fibre arrays and the PIC can be seen in figures 3.7(c) and (d).

For the study reported in [48], the insertion losses for the SMF–SMF interface (see figure 3.7(a)) were about 0.5 dB, corresponding to an efficiency of 90% at wavelengths between 1500 and 1600 nm. Regarding the insertion loss for the InP laser–SMF interface (see figure 3.7(b)), the authors reported a value of 0.8 dB, corresponding to an efficiency of 84%.

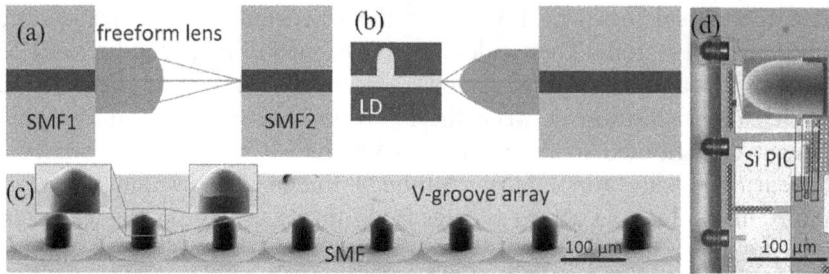

Figure 3.7. Microlenses fabricated through the TPP process. (a) Lensed SMF used to couple to another SMF. (b) Coupling of a laser diode with a lensed SMF. (c) Fibre V-groove array with eight different microlenses at the tip of each SMF. (d) Microlenses inscribed directly on a Si PIC. Reprinted with permission from [48] © 2016 Optical Society of America.

The viability of lenses directly fabricated on the edges of optical chips used in a SiP optical coherence tomography system [49] was also tested. These consisted of three different microlenses, as shown in figure 3.7(d). The couplings revealed insertion loss values of 2.8, 3.6, and 4.2 dB. These were higher than those of the SMF–SMF interface and the InP–SMF interface. This could be explained by the non-optimal design of the lenses, necessitating an optimisation of the mode-field diameter through the microlens design. Despite the losses, these lenses still have the potential to be used in SMF–PIC interfaces with low insertion losses.

More recently, the additive manufacture of beam-shaping elements capable of decreasing the coupling losses between PICs and optical fibres has been expanded, with several devices being developed. Dietrich *et al* [50] made an important contribution to the field by describing a TPP 3D-printed toolbox of optical elements fabricated on top of optical fibres or PIC waveguides. Examples of these elements included free-form lenses, free-form mirrors, lenses with high numerical aperture (NA), beam expanders, and multi-lens beam expanders. A representation of all these components and their corresponding SEM images can be observed in figures 3.8(a) and (b)–(e), respectively.

Following the TPP 3D printing of the different beam-shaping free-form elements shown in figure 3.8, the authors tested the insertion losses and efficiencies [50] of the couplings. For the printed free-form lens attached to the laser facet (seen in figure 3.8 (b)) and coupled to SMFs, they measured an insertion loss of 1 dB, which was similar to the theoretically predicted value of 0.6 dB [50]. For beam-shaping lenses attached to the SMF facet, the authors reported a laser facet coupling loss of 0.6 dB, which was close to the theoretical value of 0.2 dB. This was equivalent to a coupling efficiency of 88%, a better value than those obtained for the most optimised lensed fibres [45].

In a second set of experiments, the authors tested in situ printed curved mirrors to simultaneously define the propagation direction and beam shape, allowing surface-emitting and edge-emitting devices to be flexibly combined in compact arrangements [50]. Measurements made for mirrors printed at the facet of an InP laser (see figure 3.8(c)) showed a coupling loss of 2.9 dB (corresponding to an efficiency of

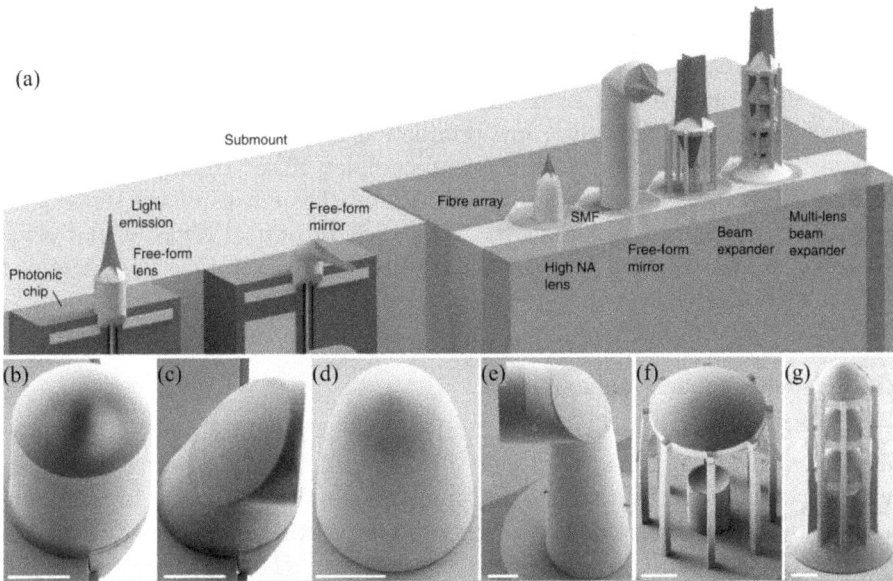

Figure 3.8. (a) Representative schematic showing different beam-shaping elements. SEM images of: (b) a free-form lens, (c) a TIR mirror, (d) a high-NA free-form lens, (e) a TIR mirror for beam deflection, (f) a beam expander, and (g) a multi-lens beam expander. Panels (f) and (g) were designed to relax alignment tolerances during the assembly process. All the SEM images have a scale of 20 μm. Reproduced from [50], with permission from Springer Nature.

59%). The high loss value was attributed to the large laser divergence at the laser facet, resulting in incomplete TIR at the mirror surface. Despite this, the losses were still lower than those reported for horizontal-cavity surface-emitting lasers complemented by monolithically integrated lenses [51, 52].

Finally, in order to relax the alignment tolerances between the elements, the authors performed a third set of experiments using beam expanders printed on the laser facet and waveguides. These consisted of multi-lens beam expanders, as shown in figure 3.8(g). These were tested for the coupling between edge-emitting lasers to SMF and passive TriPleX chips, resulting in coupling losses of 0.8 and 2.5 dB, respectively, with a 1 dB alignment tolerance of approximately ±5 μm in all directions. To achieve more compact dimensions, a concave diverging lens was used in series with a plano-convex focusing lens, as shown in figure 3.8(f). The same tolerances were also attained for more compact multi-lens systems, such as the one shown in figure 3.8(f), consisting of just one concave lens in series with a plano-concave lens [50].

3.4 Optical fibre manufacture

Optical fibres are the most widely recognised waveguides globally, forming the backbone of our communication infrastructure and playing important roles in various applications, such as sensors, imaging, and lighting. While traditional manufacturing technologies based on chemical vapour deposition (CVD) and capillary stacking have been standardised for the preform fabrication of silica fibres, they pose limitations in the use of complex designs and advanced materials. Over the past decade, AM has been considered a possible technology capable of competing with established fibre fabrication technologies. The reasons are obvious: AM offers high design flexibility, low production time, and can reduce the labour commonly required for the fabrication of complex fibre structures that rely on porous structures, which are normally fabricated through capillary stacking and drilling. The advantages of AM in the fabrication of optical fibres extend beyond these opportunities, as AM opens up the possibility of developing fibre designs for novel applications that were previously impossible.

3.4.1 Step-index polymer optical fibres

Despite being almost the same age, polymer optical fibres (POFs) have been superseded due to the transparency of silica optical fibres. However, POFs have several advantages over silica fibres, including higher elastic limits, a lower Young's modulus, greater flexibility, lower processing temperatures, lightweight properties, non-brittle characteristics, and biological compatibility. Thus, POFs have also been the subject of study over the years in specific application areas such as short-range communications, lighting, sensing, and medical instruments. These POFs comprise a cylindrical core and cladding, featuring either a graded index or step-index (SI) refractive index profile. They can be designed to exhibit high NAs and core and cladding diameters that can reach several hundred micrometres, making them generally MM. These fibres have mainly been fabricated through a discontinuous process involving the fabrication of a preform (a scaled-up version of the fibre, i.e. larger in diameter and shorter in length), followed by a heat-drawing process carried out in a drawing tower.

Preform fabrication is typically carried out in one of the following ways: (i) by filling a polymethyl methacrylate (PMMA) hollow cylinder with a mixture of monomers in varying proportions [53]; (ii) using the 'Teflon technique' [54], where the cladding material is first polymerised in a glass tube containing a Teflon rod at its centre, which is then removed and replaced with the core monomers; (iii) using the 'pull-through' technique [55], where the cladding preform is made as for the 'Teflon technique', but the core preform is polymerised in a tube that is later heat-drawn and tightly inserted into the hollow cladding preform; and (iv) using heat casting [56], where the cladding material granules are cast into a solid rod that is then drilled to allow for the injection of the molten polymer core. Despite being standardised techniques, the development of special core–cladding designs or the possibility of supporting SM behaviour are characteristics commonly required in optical fibre technologies and are hard to achieve through the aforementioned methods.

Due to the design flexibility of AM technology and the possibility of printing with several polymer materials, the use of this technology in optical fibre fabrication was soon realised [57, 58]. The first example of a polymer waveguide appeared more than a decade ago, with the direct printing of short light pipes by vat polymerisation printers [59, 60]. For extended lengths, thermosets—the raw material of these printers, which is polymerised through ultraviolet (UV) curing—cannot be employed. Instead, thermoplastics (moulded at high temperatures) are more suitable. Obviously, the direct extrusion of the polymer material through thin aperture nozzles, as is done in fused deposition modelling (FDM) 3D printers, can itself produce light pipes that effectively guide light through TIR between the extruded polymer and the surrounding air. This would allow the production of POFs with extended lengths. One such study was reported in [61], where the fabrication of circular waveguides, generally known as coreless or no-core fibres, took place through the direct FDM extrusion of fluorinated acrylonitrile butadiene (ABS) and fluorinated polyethylene terephthalate glycol (PETG) filaments. While the team was able to show thin fibre diameters reaching $\approx$30 μm and lengths exceeding 100 m, their losses were high (0.26 dB cm^{-1} @ 543 nm). However, even considering low propagation losses, these fibres do not possess an external protection layer. Thus, oil, water, dust, debris, or even the grease from fingers (i.e. when touching the fibre) can alter the light-guiding conditions, making them highly susceptible to environmental factors and thus resulting in low reliability. To overcome this, the use of a cladding material is necessary. A possible way to address this through 3D printing was demonstrated in [62], where the materials of both core and cladding regions were co-extruded in the desired core–shell configuration (see figure 3.9) through a custom-designed printhead consisting of two coaxially aligned cylindrical nozzles.

The optical waveguides made from the photocurable liquid core–fugitive shell reported in [62] were made through the encapsulation of a liquid resin (OrmoClear from Micro Resist Technology) in a viscoelastic shell composed of an aqueous triblock copolymer solution, Pluronic F127 (BASF). The fugitive shell served as a sacrificial support for the core fluid before it was UV cured immediately after the nozzle extruder. The dimensions of the extruder nozzle defined the cross-sectional dimensions of these waveguides. Through this innovative research, the authors were able to report waveguides with dimensions of 400–800 μm, featuring customisable shapes and lengths, and with minimum losses of 0.1 dB cm^{-1} in the visible region.

While the direct fabrication of fibres through 3D printers appears to be an interesting opportunity for advancing POF technology, the methodology seems more suitable for fabricating waveguides with large core diameters, which are highly MM in nature and therefore limit the range of applications in optics and photonics.

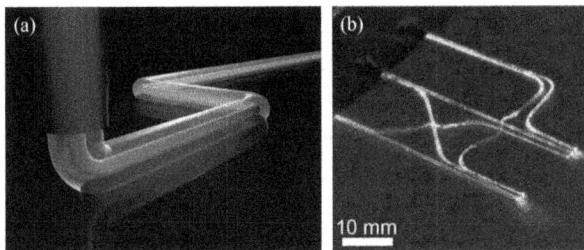

Figure 3.9. (a) Schematic of the direct printing of polymeric optical waveguides. (b) Waveguides crossing each other with minimal crosstalk [62] John Wiley & Sons. Copyright © 2011 WILEY-VCH Verlag GmbH & Co. KGaA, Weinheim.

To this end, and despite its higher cost, the traditional preform heat-drawing process appears more suitable. In this case, the research community has demonstrated the feasibility of using FDM printers for the direct fabrication of the fibre preform. Earlier studies on this topic were reported by Cook *et al* [63], who described the fabrication of an MM-SI fibre using a dual-head FDM printer. Through the use of two heads, the authors were able to simultaneously print both the core and the cladding of the fibre preform. This study used modified PETG filament for the cladding and modified ABS for the core, resulting in a refractive-index difference of 0.02. The preform was fabricated with a core diameter of 8.0 mm and a cladding diameter of 18.6 mm, and had a total length of 90 mm. After heat drawing, the authors reported a SI fibre with an elliptical shape, having outer dimensions of approximately 283 µm and 204 µm, and core dimensions of roughly 171 µm and 60 µm. The nonsymmetrical fibre profile was nonideal and resulted from the chosen print orientation, in which the fibre's principal axis was aligned parallel to the printing bed, thereby affecting the fibre's cross-section. This effect was further exacerbated during the fibre-drawing process. Despite the low refractive index difference, the final diameter of the fibre led to highly MM behaviour. Furthermore, loss measurements revealed values of 0.6 and 0.9 dB cm^{-1} for the visible and telecom regions, respectively, which were mainly attributed to scattering losses between the core and cladding regions [63]. While the use of a dual-head extruder 3D printer is less conventional, the use of standard single-head FDM 3D printers can also be an option for fabricating fibre preforms and for fibre drawing. An example of this capability was described by Gozzard *et al* [64]. The authors of this paper had to change the core and cladding materials twice during the printing process, making it more laborious. The fabrication method was identical to that described in [65]. The preform was printed with its length parallel to the printing bed. The outer shape of the preform was hexagonal, with a distance of 2.75 mm between opposite corners. For the printing process, a nozzle with a diameter of 0.2 mm was used. During the preform printing process, the printing was stopped when half of the PETG cladding was complete. The filament was then changed to ABS, and a single 0.2 mm ABS filament was deposited in the middle hollow region of the printed PTFE cladding (to achieve SM behaviour, the extruder nozzle was adjusted to 0.1 mm, resulting in fibres with acceptable diameters for handling). Finally, the printer nozzle was again changed to PETG, and the second half of the cladding region was deposited. The preform was later extruded through a 0.4 mm diameter nozzle in the 3D printer. To control the draw rate of the preform, a stepper-driven spindle was used, providing control of the fibre diameter. The step-by-step fabrication process of the fibre is shown in figure 3.10.

The entire process of printing the preform and completing the drawing process took approximately 20 min, and losses of between 0.8 and 1 dB cm^{-1} were reported for the 1550 nm region. While the authors could theoretically draw fibres to obtain cores smaller than 4.8 µm to support SM operation, they were unable to properly cleave the fibres to observe and prove their monomodal operation.

The design flexibility of AM introduces the possibility of exploring the fabrication of optical fibres with special core shapes and arrangements, giving another degree of

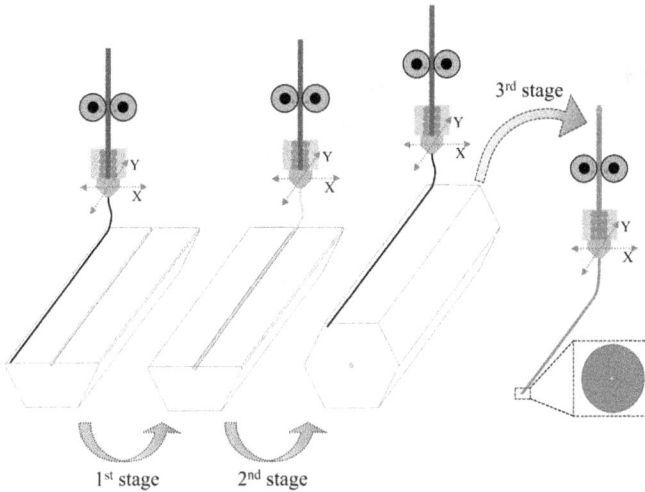

Figure 3.10. Step-by-step fabrication process of a polymer SI-SMF reported in [64]. In the first stage, the first half of the fibre cladding is printed in PETG filament through a 0.2 mm nozzle. Then, the filament is changed to ABS, and the extruder nozzle is set to 0.1 mm, allowing the printing of a single filament in the central region of the 3D-printed part. The second half of the cladding region is then printed under the same conditions as those used for the first half. Finally, the fibre is heat-drawn through a 0.4 mm extruder nozzle in a 3D printer.

freedom in the control and manipulation of light propagation, such as control of the mode-field diameter, light polarisation, etc. Examples of such design flexibility have been reported for the fabrication of SI fibres with unconventional core shapes, such as square, triangular, and rectangular [66]. An example of this achievement is shown in figure 3.11 for fibres composed of polycarbonate (PC) cores and ABS copolymer claddings.

To obtain the results reported in figure 3.11, the authors 3D printed the preform cladding region with an inner hole that complemented the specially shaped core for each specific fibre type. After printing, ready-made PC pipes with cross-sectional shapes polished to the desired shape were inserted into the hollow region of the 3D-printed preforms and then heat-drawn into fibres in a drawing tower. The drawn fibres had diameters reaching $\approx$500 μm and cores reaching tens of microns [66]. Given the refractive indices of the core and cladding materials (i.e. 1.5593 and 1.5812), the parameter V (i.e. $V = (2\pi a/\lambda)\ (n_{co} - n_{cl})^{1/2}$, with a as the core radius, and n_{co} and n_{cl} as the core and cladding refractive indices, respectively) is higher than 2.405; thus, the fibre behaves as an MM waveguide. Losses measured at 633 nm in the produced fibres reached values in the range of 1–2 dB cm^{-1}.

3.4.2 Microstructured polymer optical fibres

The light-guiding mechanism in SI fibres relies on the use of different refractive indices for the core and the cladding material. However, when SM behaviour is required, a compromise must be made between the refractive index contrast and the core diameter to obtain a V-parameter $\leqslant$2.405. In other words, one can choose

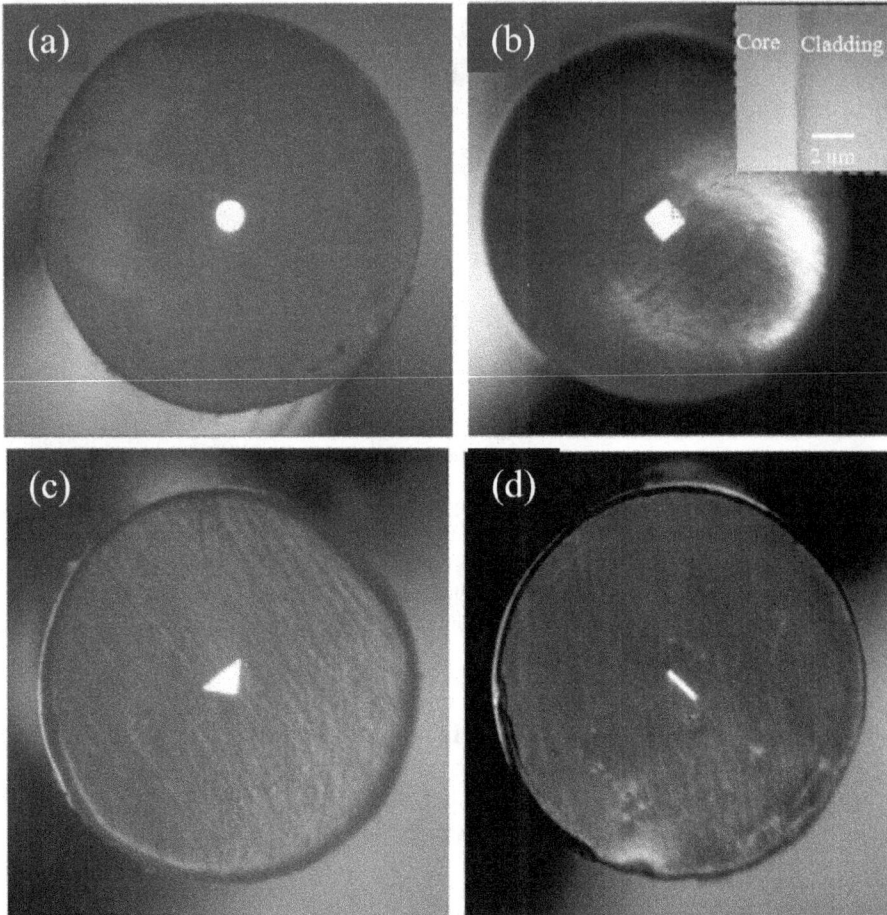

Figure 3.11. Polymer SI fibres with special core shapes, obtained using a technique consisting of preform 3D printing and heat-drawing: (a) circular, (b) square, (c) triangular, and (d) rectangular-core fibres. The cladding of the fibres is about 500 μm. Reprinted from [66], Copyright (2017), with permission from Elsevier.

between small cores and a high core–cladding refractive index contrast or larger cores and a modest refractive index contrast. Despite the opportunity to employ a wide range of polymer materials for the 3D printing of polymer preforms or fibres, their high refractive index contrast necessitates the use of small cores. Furthermore, considering that the transparent wavelength region of polymers is located in the visible region, this forces the core radius to become even smaller. This surely poses challenges for light coupling in and out of the fibre; thus, solutions based on microstructured optical fibres, also known as photonic crystal fibres (PCFs) [67], have been proposed. These are single-material optical fibres where an array of air holes is arranged in the cladding region to allow light guidance in a similar way to the TIR mechanism. Through proper design of the air-filling fraction (i.e. d/Λ, where d is the hole diameter and Λ is the hole-to-hole distance), one can obtain fibres with unlimited SM behaviour; in other words, the ability to support SM transmission at all wavelengths for which the material is transparent. Additionally, the mode field in these fibres can be concentrated in a small area, allowing for high intensity within the fibre, which boosts highly nonlinear effects for supercontinuum generation. Furthermore, other possibilities include the use of asymmetric structures to produce fibres with high birefringence or the adjustment of the air-filling fraction to control fibre dispersion. Based on these opportunities, and considering the low processing temperatures of polymers, the research community has already explored the fabrication of POFs through different methods, including stacking, drilling, extrusion, and casting [68]. The fabrication complexity, combined with the limited design flexibility offered by these techniques, led the scientific community to explore AM technology. Cook *et al* were among the first to report the 3D printing of these air-structured POFs a decade ago [69]. Later, several subsequent studies were completed. Some examples are shown in figure 3.12.

In the study reported by Cook and co-workers [69], the authors used an FDM printer to directly print an ABS preform 10 cm tall and 1.6 cm thick, containing a solid core and six air holes ($d = 200$ μm) equally spaced in the fibre cladding region (see figure 3.12(a)). The preform was later heat-drawn into $\approx$10 m of fibre (without air-hole pressurisation). Cross-sectional images of the fibre revealed a slightly elliptical shape, with cladding dimensions measuring approximately 712 μm and 605 μm, and core dimensions of roughly 221 μm and 148 μm. The preliminary results showed that the fibre supported MM propagation, presenting losses of 1–2 dB cm^{-1} in both the visible and telecom wavelengths. Despite this, the fibre was unable to perform as a waveguide within the solid-core inner region [69], probably due to the high air-filling fraction, indicating that further improvement was still needed.

Material loss plays a crucial role in optical fibres. Due to the high attenuation of polymers, this issue is even more critical for POFs. Additionally, the effects of other material properties, such as nonlinearities, can influence the fibre's guiding properties. Hollow-core (HC) microstructured optical fibres present an intriguing opportunity due to their air-guidance mechanism, which helps address the issues associated with polymer attenuation. The potential to merge the advantages of both AM and HC fibres has generated interest within the research community, and different authors have already dedicated their time to this topic [70, 71, 73, 75]. The

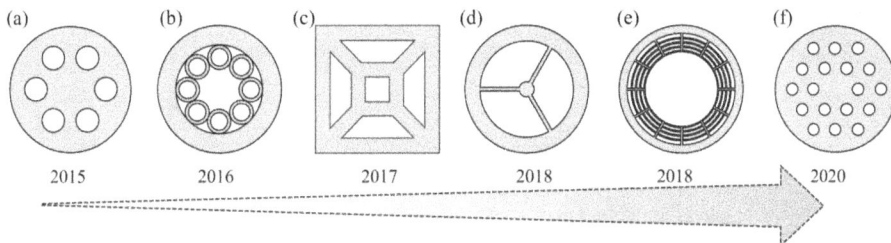

Figure 3.12. Examples of 3D-printed microstructured POFs produced over the years: (a) one-layer microstructured fibre [69], (b) circular hollow-core (HC) fibre [70, 71], (c) square HC fibre [71], (d) suspended core fibre [72], (e) anti-resonant fibre [73], and (f) two-layer microstructured fibre [74].

first attempt was made by Zubel *et al* [70], inspired by prior studies of silica air-guiding fibres and 3D-printed polymer waveguides for terahertz applications, which relied on the negative curvature of the core wall and the scattering characteristics of the cladding elements (see figure 3.12(b)). Despite the early report, Zubel and co-workers only showed the printing of a PMMA preform, followed by its drawing into a 4.7 mm cane. Later, Marques *et al* [71] reported the 3D printing and fibre drawing of the same circular HC fibre structure reported earlier, and also a rectangular-core HC fibre (see figures 3.12(b) and (c), respectively). The fibres were printed in ABS, and the subsequent steps followed the same procedure as that used to draw regular POFs, namely, initially reducing a 50–60 mm fibre preform to a 25 mm cane, which was later further reduced to 12 mm for the final fibre drawing. This process produced circular and rectangular HC fibres with external diameters of 170 and 250 μm, respectively. Light injected from a supercontinuum source into a rectangular-core fibre 26 cm long elucidated an air-guiding mechanism. However, the result was still too preliminary, and the team continued their investigations. A year later, using another HC fibre design, the team—motivated by the potential to guide light in the fibre's HC region through Bragg reflection and to exploit the strong interaction between light propagating in the HC region and substances such as fluids and gases for sensing applications—successfully demonstrated the air-guiding mechanism in the mid-IR region [73]. The fibre consisted of a circular-shaped core surrounded by four periodic layers of polymer cladding separated by air gaps, as illustrated in figure 3.12(e). After 3D printing and fibre drawing, an HC fibre with an external diameter of 466 μm was obtained, as shown in figure 3.13(a).

Upon comparing the HC fibre designed and schematised in figure 3.12 and the one shown in figure 3.13(a), the authors realised that severe deformation had occurred during the drawing process. However, modal images taken at wavelengths of 3.5–5 μm by a thermal infrared camera revealed that light was confined to the air-core region even when the fibre was bent. Modelling results for the drawn fibre indicated that light was restricted to anti-resonant (AR) reflection at the first layer of polymer strands. Characterisations of the propagation loss revealed values on the order of 30 dB m^{-1}, which were two orders of magnitude smaller than the PETG loss, but still left room for improvement through the reduction of structural deformations occurring during printing and fibre drawing.

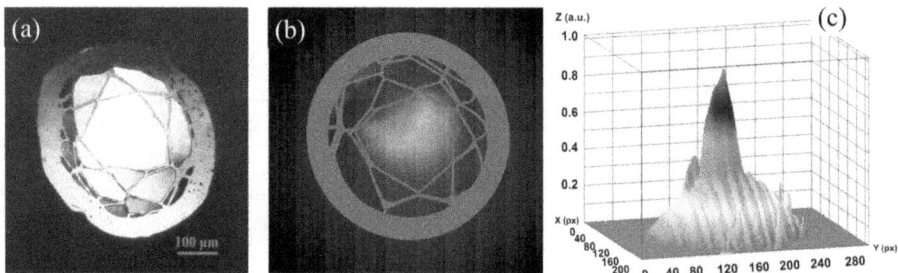

Figure 3.13. (a) Cross-sectional microscope image, (b) mode image in the mid-IR, and (c) near-field intensity of the HC fibre. Reproduced from [73]. CC BY 4.0.

The fabrication of POFs aided by 3D printing often involves a two-step procedure, namely 3D printing followed by fibre drawing. This is a time-dependent process and requires specialised drawing towers, which are only available in a few research labs around the world. The research literature has already described the production of SI fibres directly from 3D printers, as detailed in figures 3.9 and 3.10 from studies reported in [62, 64], respectively. While numerous studies have reported the capability to fabricate fibres with microstructured designs through a single-step 3D printing procedure, their operational wavelengths are located in the terahertz region [58], allowing printing dimensions to match the guidance requirements of those fibres. However, for shorter wavelengths, fabricating fibres with microstructured designs through a single-step 3D printing procedure is more challenging. Despite this, the use of 3D printers with custom-made extruder nozzles capable of printing microstructured fibres [72], as illustrated in figure 3.12(e), has already been demonstrated. Even with the direct fabrication approach, this study relied on a machined extruder nozzle intentionally designed for the shape of the fibre (see figures 3.14(a)–(c)); thus, the design flexibility of 3D printing was not the focus of the research. Instead, the team focused on the capability to fabricate fibres through an alternative method, without the necessity of using fibre-drawing tower machines. The work relied on the fabrication of a suspended core fibre made of ABS. After fibre extrusion, the team collected microscope images and near-field images in the 1550 nm region, as shown in figures 3.14(d)–(f), respectively [72].

The extruded fibre had a diameter of 800 μm and exhibited MM behaviour, a condition that could be suppressed by adjusting the fibre parameters before fabrication or by pressurising the air holes during the drawing stage. Propagation losses measured in the 1550 nm region reached 1.1 dB cm^{-1}, which was primarily attributed to fibre surface roughness. The authors also characterised the fibre for bending loss, measuring a loss of 20 dB m^{-1} for a bending radius of 51.7 mm.

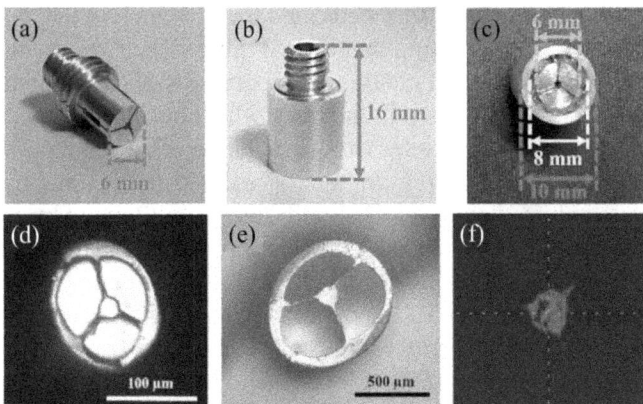

Figure 3.14. (a) Structured nozzle body and (b) cover. (d) Structured nozzle after drawing of the fibre. (d) and (e) Microscope images of microstructured optical fibres extruded with different diameters. (f) Near-field image obtained at a wavelength of 1550 nm with a fibre bending radius of 12.5 mm. Reproduced from [72]. CC BY 4.0.

3.4.3 Silica optical fibres

The low melting temperatures of polymers, i.e. 200 °C–300 °C, make them easy to process, which led academia to first report the use of AM for fabricating POF preforms through FDM printers back in 2015 [69]. This marked a new transitional era for the fibre fabrication process, offering the opportunity to develop fibre structures previously thought impossible and opening the door to new applications. However, silica undoubtedly has higher transparency, making it the most preferable host material for fibre fabrication. Furthermore, silica outperforms polymers in several properties, including hardness, thermal resistance, chemical resistance, and chemical tunability. Yet silica has a high melting temperature, i.e. 1900 °C, making it challenging to fabricate fibres from silica through regular extrusion processes. The printing challenge becomes even greater when one considers the high viscosity of silica. However, to circumvent this, researchers have dedicated their efforts to developing hybrid materials composed of mixtures of organic and inorganic components that can be directly printed and later sintered to remove any traces of organic materials. Some of the first studies reporting the free-form fabrication of transparent silica glass-structured nanocomposites were published in 2017 and 2019. In these studies, the nanocomposites were made of amorphous silica nanoparticles dispersed in UV-curable monomers. The 2017 study used micro-SLA [76], and the 2019 study used TPP [77]. The fabricated structures achieved high-resolution features but were limited to millimetre sizes. Direct ink writing (DIW) [78, 79] has also been proposed. This consists of the printing of colloidal silica suspensions to form silica green bodies that are later sintered to a fully dense structure. Through this method, silica structures have achieved high transparency at the centimetre scale. Taking into account these achievements, it became clear that the free-form fabrication of silica optical fibres was no longer impossible, and Chu *et al* [80] were among the first to report the preform fabrication and fibre drawing of a silica SMF.

Chu *et al* [80] were interested in providing a proof of concept for the use of 3D printing in fibre fabrication. Thus, they dedicated their research to the most widely used and simplest type of optical fibre, namely, SI fibre. Their work began with the preparation of photocurable resin containing amorphous silica nanoparticles with an average diameter of 40 nm dispersed in the resin. The fabrication was performed using a digital light processing (DLP) printer with a resolution of 75 μm, focusing only on the cladding region of the fibre. Then, a mixture of uncured resin containing germanium dioxide (GeO_2), titanium dioxide (TiO_2), and 2,2-Azobis(2-methylpropionitrile) was poured into the central hollow region of the free-form cladding and left to polymerise at temperature. Next, the debinding process took place at different increasing temperature steps, allowing the removal of the organic material and aggregation of the silica nanoparticles. Finally, the preform was installed in a quartz tube for support and drawn into ≈2.3 km of fibre at 1850 °C. The step-by-step fabrication process can be seen in figure 3.15 [80].

The refractive index profile of the sintered preform was analysed, showing that the refractive index difference between the inner and outer clad regions was quite small, i.e. 2.3×10^{-4}. The authors were able to report MM fibres with circular

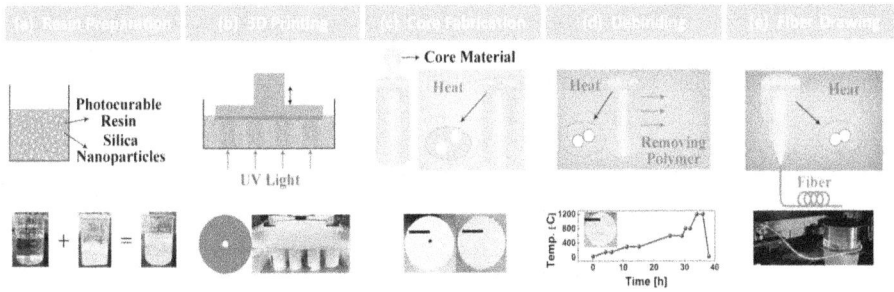

Figure 3.15. Step-by-step process for the fabrication of a silica SI fibre from a 3D-printed preform. (a) Preparation of the resin mixture with the silica nanoparticles. (b) Printing of the cladding preforms using a DLP 3D printer. (c) Insertion of the liquid core material into the hollow region of the cladding preform and thermal curing. (d) Debinding process of the cured components. (e) Drawing of the preform into a fibre. Reprinted with permission from [80] © 2019 Optical Society of America.

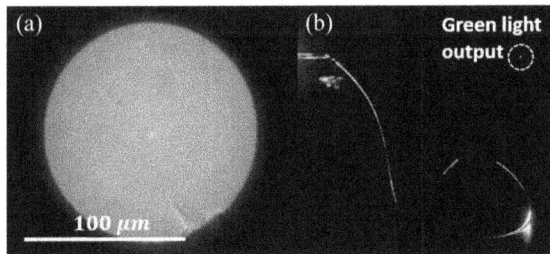

Figure 3.16. SM-SI fibre produced by preform 3D printing and heat-drawn in a drawing tower: (a) cross-sectional image of the fibre; (b) light transmission through the fabricated fibre at 532 nm. Reprinted with permission from [80] © 2019 Optical Society of America.

cladding and core shapes with diameters of 242 and 14 μm, respectively, and SM fibres with circular cladding measuring 131 μm and an elliptical core shape with dimensions of 3.6 and 4.8 μm, concluding that a more uniform tension would be required during the drawing to obtain a more homogeneous core shape. A cross-sectional image of the SM fibre, together with its light transmission, can be seen in figures 3.16(a) and (b), respectively [80].

The transmission losses measured by the cutback method in 2 m of MM fibre reached ≈ 24 and ≈ 5.5 dB m^{-1} at wavelengths of 532 and 1550 nm, respectively. However, for the SM fibre, the authors measured enhanced performance in the visible region, i.e., ≈ 13.4 dB m^{-1}. The cutoff wavelength of the SMF was located at 780 nm; thus, measurements at 1550 nm did not produce good results, resulting in a high loss of ≈ 114 dB m^{-1} [80]. Overall, the proof of concept of using 3D printers to fabricate optical fibre preforms demonstrated neither excellent transmission loss performance nor complex fibre designs. However, it sparked the transition to a new fabrication methodology, and subsequent studies have followed.

While the use of the DLP technique has been one of the most widely reported for the fabrication of silica fibre preforms [58], other techniques have also been reported, such as DIW [81], selective laser sintering (SLS) [82, 83], and laser powder deposition (LPD) [84, 85]. FDM of soft glasses with low melting temperatures, such as soda-lime glass, chalcogenide glass [86], and heavy metal oxide glass [87], has also been applied for the fabrication of silica optical fibres. Examples of these are shown in figure 3.17.

The demand for special optical fibres capable of creating different opportunities in both telecom and sensor applications has also driven researchers to develop complex silica fibre structures that extend beyond SI-SM and SI-MM configurations. Some examples include AR fibres [82], PCFs [82, 88, 89], dual-core fibres [90, 91], and MCF [90, 92, 93]. Some 3D-printed fibres are shown in figure 3.18.

The material selection is always important when manufacturing an optical fibre. This selection is primarily performed to ensure the refractive index contrast required for light guidance. However, the material choice is also important when special features need to be added to the fibre, such as amplification, fluorescence, lasing, sensing, etc. While popular CVD methods can be used to fabricate doped preforms, it is simpler to mix dopants with SiO_2 nanoparticles at varying ratios, and such mixtures can readily be incorporated into specialised 3D-printed cladding preforms

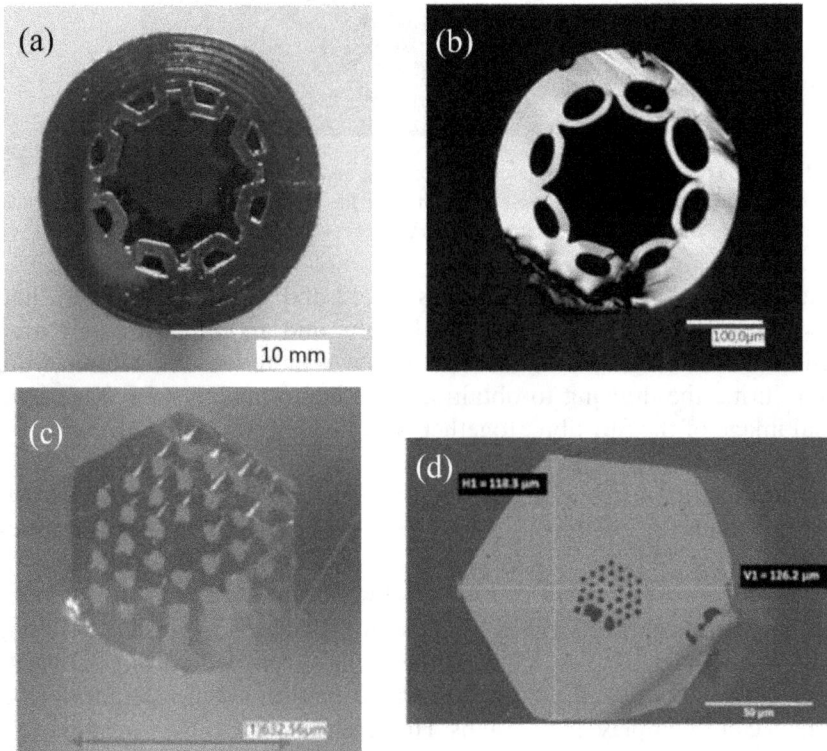

Figure 3.17. Fabrication of soft glass silica fibres: (a), (b) chalcogenide glass HC fibre [86], and (c), (d) heavy metal oxide glass PCF. The images in (a) and (c) are the preforms, and the images in (b) and (d) are the drawn fibres. Reproduced from [87]. CC BY 4.0.

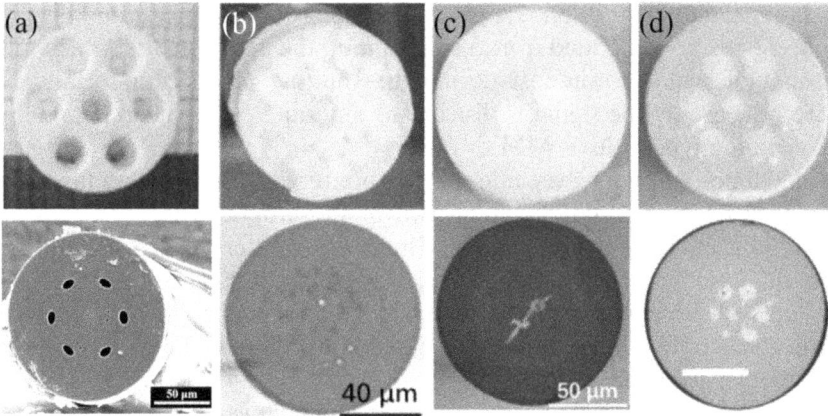

Figure 3.18. The top images show silica preforms, and the bottom images correspond to silica drawn fibres from different references. (a) Air-hole Yb-doped silica core PCF. Reprinted from [88], Copyright (2021), with permission from Elsevier. (b) All-solid PCF. Reproduced from [89]. CC BY 4.0. (c) Dual-core fibre. Reprinted from [91], Copyright (2024), with permission from Elsevier. (d) MCF. Reproduced from [92]. CC BY 4.0.

in liquid or powder form. The use of this method has led to a report of the fabrication of an air-hole PCF, where the central preform region was filled with ytterbium (Yb)-doped silica [88] (see figure 3.18(a)); an all-solid PCF, where the holes were filled with a powder of borate (B_2O_3) and SiO_2 nanoparticles [89] (see figure 3.18(b)); dual-core and MCFs that included bismuth (Bi) and erbium (Er) as dopants [88] (see figures 3.18(c) and (d)); highly birefringent MCFs incorporating TiO_2, GeO_2, B_2O_3, and aluminium oxide (Al_2O_3) [93], etc.

While the fabrication of fibres through 3D printing has been the subject of much research during the last decade, the most common conclusions of these studies were that the produced fibres still have low transparency, are challenging to operate as SM fibres, tend to show fibre geometrical defects during the drawing process, tend to show high scattering at core–cladding boundary transitions, etc. The technology is still evolving, and its advantages over other established techniques will surely lead to more research focused on optimising fabrication to compensate for the reported problems.

3.5 Waveguide splitters

Nowadays, the need to combine and split optical waves is essential for several photonic applications, including interferometry, PICs, quantum–optical experiments [94, 95], and fibre optic networks, particularly in passive optical networks (PONs), where the efficient and cost-effective distribution of high-bandwidth services is achieved in modern fibre optic communication systems. Therefore, splitters play an important role in this field. These are passive optical components that can split or combine optical signals into multiple input or output arms with minimal insertion losses. They achieve this signal division by physically splitting the light within the optical waveguide from one input into multiple outputs. Key considerations in designing splitters include the split output signal ratio; the number of output signals, N, generated from a single input; the insertion loss, which refers to the amount of signal power lost during the splitting process; uniformity, which relates to how evenly the signal is distributed among the output ports; and finally, the waveguiding type—either MM or SM.

So far, the best-known waveguide splitters are fused biconical fibre tapers. These are produced by fusing and stretching two or more optical fibres so that light is forced to split in the tapered region. These are splitters optimised for specific wavelengths. On the other hand, planar waveguide circuits are another type of splitter that has been implemented in semiconductor technology to create waveguides on a substrate. Unlike traditional fused fibre splitters, these can operate over a wider wavelength band.

The most commonly used approaches for splitting power from one branch to 'N' branches are Y-branch splitters and multimode interference (MMI) splitters. In the Y-branch approach, the splitters have one input arm and two output arms. Typically, the splitting ratio can be designed as needed, allowing for specific power outputs in each arm [96]. The advantage of this approach is that these splitters are polarisation- and wavelength-independent. The second category of splitters, the MMI splitters, is based on the self-imaging effect [97]. The advantages of this approach include low losses, low polarisation dependence, wide manufacturing tolerances, and more even power splitting. However, care must be taken, as they are dependent on wavelength.

Large-core MM waveguides with diameters extending to the millimetre range have several applications, such as lighting or decoration, short-range communications, and automotive applications, particularly for multimedia data transmission, including multimedia-oriented system transport (MOST), IDB 1394, and FlexRay communication systems. To date, several methods have been developed to manufacture splitters for these MM waveguides. Examples of these methods include the hot embossing method [98, 99], the side polishing method [100], injection moulding in a metallic moulding mask [101], and even the light-induced self-written (LISW) waveguide technique, together with wavelength division multiplexing filters [102]. The manufacture of planar optical splitters through different methods, such as ion exchange in glass [103] or the photolithographic process [104], is a feasible and trustworthy approach. Mass production of these components is essential, and methods based on laser Lithographie, Galvanoformung, and Abformung (LIGA)-technique moulding [105], injection moulding [106], and the hot embossing method using thermoplastic resin [98], among others [107, 108], have already been tested. Despite this wide array of techniques, none have been able to establish themselves as commonly used methods in the manufacture of splitters. This may be due to several factors, including the complexity of certain technologies that render them difficult to implement, as well as challenges related to mass-producing components. In this context, AM appears as an alternative. This technique has been employed in this context for a long time and was first demonstrated for planar waveguides using inkjet technology [109–111].

3.5.1 Planar waveguide splitters

Inkjet printing offers precise volumetric control of dispensed material, with low cost and speed as its key features. This method has been used with both thermoplastic and thermosetting polymers for the fabrication of micro-optical components utilising a drop-on-demand piezoelectric printhead. In this system, a droplet is only ejected when a voltage pulse is applied to a transducer coupled to the fluid dispenser, allowing temporal and spatial stability. The dispensed droplets can be released at a rate of up to 8 kHz while the substrate is continuously rastered back and forth. While the droplets achieve a spherical shape when in contact with the substrate, the aspect ratio of the printed optical micro-optical elements is controlled by adjusting the viscosity of the liquid polymer being printed and by the substrate's wettability, which can be tailored through the application of coatings with appropriate wetting characteristics. Between subsequent drop depositions, in situ curing flashes enable the polymerisation of the material, allowing the continuous fabrication process to continue [111]. This versatile technology has demonstrated its capabilities in the production of planar bifurcated MM splitters, as illustrated in figure 3.19 for a 1 × 32 splitter fabricated by depositing a photopolymerisable resin onto a glass substrate with a lower refractive index.

The channel width of the splitter presented in figure 3.19 is about 100 µm. These dimensions depend on the droplet size, which is controlled by the diameter of the orifice in the dispensing device, as well as the viscosity of the resin. The control of these parameters enables the generation of droplets with typical diameters ranging from 20 to 60 µm. Additionally, to ensure the growth of the waveguide, the spacing between droplets must be small enough to allow for the coalescence of adjacent deposits but not so large as to create discontinuities in the printed lines. If the parameters mentioned above are carefully optimised, the production of MM waveguide splitters via inkjet printing is a viable manufacturing alternative to the established planar waveguide technologies based on photolithography and etching. This opportunity also applies to other AM technologies, such as the maskless laser direct writing technique, which has been used for the direct writing of 1 × N MM planar splitter waveguides on ultraviolet photocurable polymers [112].

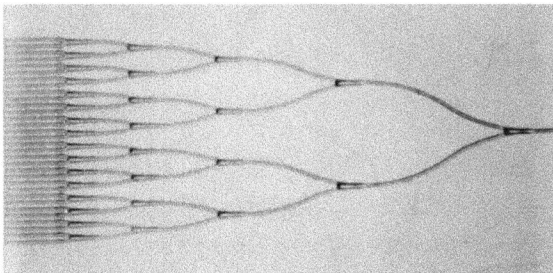

Figure 3.19. A 1 × 32 symmetric optical power splitter fabricated by inkjet printing a photocurable polymer onto a glass substrate. The waveguide is MM with a width of 100 µm. Reprinted with permission from [109]. Copyright (2001) American Chemical Society.

While inkjet printers and UV direct laser printing offer design freedom by enabling the use of curved shapes in splitter designs [112], they also allow for more compact devices, as curves are easier to produce than branches made of straight lines. In line with simplicity and production scalability, Prajzler *et al* [113] demonstrated the use of AM technology for the fabrication of a large-core MM 1 × 2 Y-splitter based on a casting methodology. To achieve this, a negative mould with the shape of the splitter design was 3D printed using masked stereolithography (MSLA). A representative schematic of the proposed splitter is shown in figure 3.20(a).

The splitter described in [113] was designed to operate in the visible region and to equally split the light input signal (P_{in}) from a large-core MM fibre (500 μm core and 550 μm cladding) into two output arms (P_{out1}, P_{out2}) with cross-sectional dimensions that match the fibre diameter. The negative mould was fabricated with a U-groove shape, featuring an angle of 1.1° that separates the two output arms, a tapered length of $d = 19.2$ mm, and an S-bend region of $L_s = 22.2$ mm. These dimensions were simulated and optimised for the 650 nm wavelength region, specifically to achieve a 50/50 split ratio, considering refractive indices of 1.4109 for the cladding (polydimethylsiloxane Sylgard 184, Dow Corning) and 1.4277 for the core (LS-6943, NuSil), respectively.

After the AM of the negative 3D-printed mould, elastomer Sylgard 184 in its liquid form was poured onto it and cured at a specific temperature. The hardened part of the elastomer was then released from the mould and acted as the substrate of the waveguide. Images of the support cladding material after it was removed from the negative mould can be seen in figure 3.20(b). Later, the fibres were positioned on the terminals of the U-trenches present on the substrate and then filled with liquid LS-6943, NuSil, which acted as the core. The hardening process was subsequently followed by heat treatment, and a final layer of Sylgard 184 was used to cover the core region.

The insertion losses of the splitter were measured for the projected wavelength, yielding a loss of approximately 1.5 dB and a uniformity of 0.9 dB. These results represent promising opportunities in terms of performance when compared with

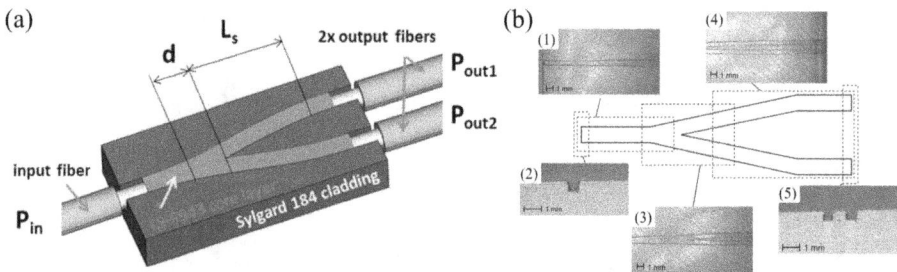

Figure 3.20. (a) Schematic of a 1 × 2 MM fibre splitter composed of a core waveguide channel in a support cladding material. The splitter is fabricated using a casting methodology based on a 3D-printed negative mould. (b) Top views ((1), (3), (4)) and cross-sectional views ((2), (5)) of different regions of the support cladding material following removal from the 3D-printed negative mould. Adapted from [113], with permission from Springer Nature.

others reported in the literature for similar waveguide dimensions [101, 105, 107], indicating that AM is a viable approach that enables scalability of the fabrication process at reduced cost.

3.5.2 Tridimensional splitters

Although planar 2D splitters offer design simplicity, an optical 3D splitter, on the other hand, offers several advantages, especially for integrated photonics and optical communication systems. In a 3D splitter waveguide approach, light is routed in multiple planes, minimising the overall chip size. This contrasts with planar designs, where all waveguides need to be in a single layer, which requires more space for complex splitting arrangements. Additionally, designing a splitter in three dimensions allows for smoother transitions and adiabatic tapers, which reduce losses associated with sharp bends or junctions. The ability to split light in all directions is also attractive in free-space optical systems, photonic interposers, and 3D PICs. This enables easy coupling to free space or fibre optic systems, which can be found in different core arrangements (MCFs) and fibre bundle arrangements. Additionally, 3D design freedom enables the waveguide mode to be shaped to match those found in the interfacing systems, thereby minimising coupling losses and avoiding the limitation of rectangular cores presented above for AM techniques based on casting methodology [113] and direct UV laser inscription on a photopolymerisable resin [112], and the plano-concave cores presented for inkjet printing [111]. Three-dimensional splitter designs also offer the opportunity to utilise more efficient manufacturing techniques, such as LIGA, femtosecond laser writing, and 3D printing through SLA and TPP, enabling the creation of high-quality 3D optical structures.

Recently, Oliveira *et al* reported the fabrication of MM large-core three-dimensional splitters (1 × 2 and 1 × 4) fabricated through direct 3D printing on a low-cost DLP 3D printer [114]. A 2D representation of the splitter is shown in figure 3.21(a), while 3D views of the sliced models of the 1 × 2 and 1 × 4 splitters can be seen in figures 3.21(b) and (c).

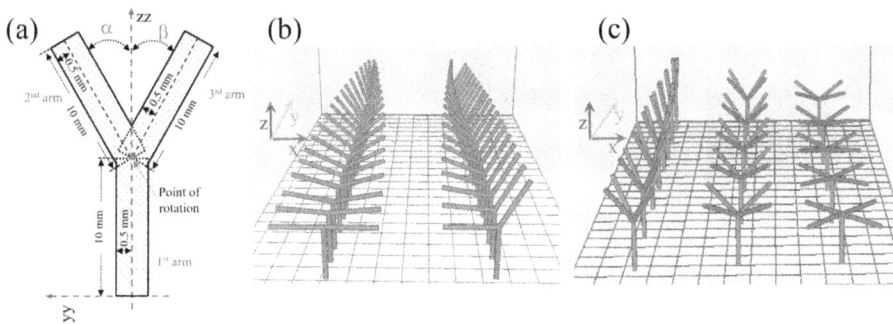

Figure 3.21. Schematic representation of a 1 × 2 MM splitter. Adapted with permission from [114]. © 2022 Optica Publishing Group.

The tridimensional MM splitters were designed to work in the visible region and to couple large-core MM fibres with a 1 mm core diameter [114]. They were designed to split power from one arm to two or four arms, each with straight lines of 10 mm length and with α and β angles separating each output arm from the principal axis of the first arm (see figure 3.21(a)). The printing was performed with 50 μm layers and an XY resolution of 50 μm, determined by the light projector. The splitters were printed vertically with the input arm secured to the building platform, as shown in figures 3.21(b) and (c). Following the printing process, the splitters were subjected to a coating step, in which they were immersed in the same resin used for printing and then hung upside down to drain the excess resin. After this hanging process, the resin was photopolymerised using UV light. This coating process enabled the removal of the stair-like effect that arose from the 50 μm layering process (see figures 3.22(d) and (f)) and the XY pixel projector dimensions (see figure 3.22(a)). Finally, the splitters were dip-coated in a UV photopolymerisable resin with a lower refractive index than that of the core. After UV curing, this external layer acted as the cladding region of the splitter. Intermediate images taken during these post-processing steps and after injecting light at the input terminal can be seen in figure 3.22.

After the whole splitter manufacturing process presented in [114], the propagation losses of a 3D-printed straight waveguide were estimated for the visible and infrared spectral regions, showing lower attenuation in the wavelength region close to ≈ 700 nm, i.e. 1.3 dB cm^{-1}. The splitter's split ratio was then measured at a nearby wavelength, specifically 627 nm, injected through the input arm and measured with a power meter at each of the output arms. The experimental and theoretical results obtained for the 1 × 2 and 1 × 4 splitter configurations for different α and β angles are shown in figure 3.23.

The theoretical results for the symmetrical 1 × 2 splitter configuration reveal that the output power is evenly distributed between the two arms, presenting minimal discrepancies between the output powers of the two arms (i.e. 45:47%, 42:43%,

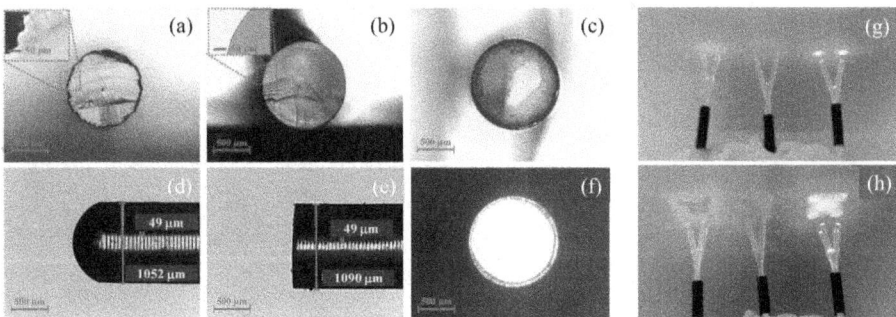

Figure 3.22. Top view of one arm of the splitter after: (a) the 3D printing of the splitters, (b) the first dip-coating process, and (c) the second dip-coating process with a low-refractive-index resin. Lateral view of the extremity of one arm of the splitter (d) before and (e) after the cleaving process. The photo in (f) presents the light output from one output splitter arm, while the photos in (g) and (h) show the final 1 × 2 and 1 × 4 splitters transmitting light at different visible wavelengths. Adapted with permission from [114]. © 2022 Optica Publishing Group.

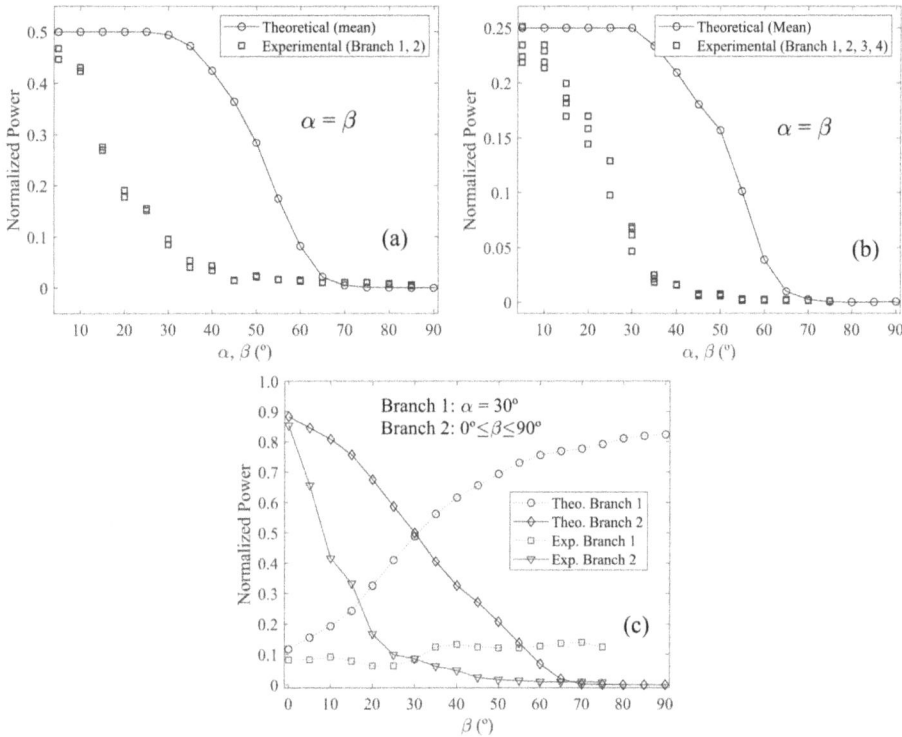

Figure 3.23. Simulated and experimental results for the symmetrical (a) 1 × 2 splitter, (b) 1 × 4 splitter, and (c) the asymmetrical 1 × 2 splitter. Reprinted with permission from [114]. © 2022 Optica Publishing Group.

27:28%, etc. for $\alpha = \beta = 5°$, 10°, and 15°, respectively). This homogeneous power distribution among the different arms was also observed for the 1 × 4 splitter. The experimental results for the symmetrical 1 × 2 and 1 × 4 splitters reveal that for angles greater than 10°, the losses become significant; i.e. for angles of 5°, 10°, 15°, and 20°, the excess loss reaches values of 9%, 15%, 46%, and 63% for the 1 × 2 splitter, and 7%, 10%, 26%, and 35% for the 1 × 4 splitter, respectively. Despite the high losses for angles exceeding 30°, the splitting ratio performance for angles equal to or less than 20° is acceptable for the symmetrical splitters. Regarding the asymmetrical splitter, it was verified that the split ratio obtained for $\alpha = 30°$ and $\beta = 0°$ matched the theoretical prediction, achieving a splitting ratio of 8:85%, which was slightly closer to the theoretical value of 12:88%. Despite these favourable results for smaller β angles, the excess loss presented was as high as 7%. When the angle β increased, the value measured in the first branch was almost constant, reaching splitting ratios of around 8%–9% for $0° \leqslant \beta \leqslant 60°$.

Despite the uniform power distribution in each of the different arms of the splitters, they presented high losses associated with the refractive index contrast across each printed layer due to the accumulated UV dosage along each printing layer. Apart from the problems associated with the stair-like effect that appears at the outer borders of the printed optical components due to pixel resolution and layer thickness, refractive index uniformity is also a crucial consideration when designing an optical device. Since UV penetrates through each of the layers that form the object and considering the high attenuation of polymers in the UV region, the layer region closest to the UV light projector will have higher UV dosages than the rest, leading to a refractive index contrast along the printing layer. This is an effect that has not yet been thoroughly discussed in the research literature. Yet, this was one possible reason for the high losses observed in the $1 \times N$ splitters described in [114]. Thus, the fabrication of 3D splitters through DLP and MSLA methods still has room for improvement.

Photonic waveguides are prime candidates for integrated and parallel ultra-fast photonic interconnects with ultra-low latency. These interconnects are crucial for next-generation optical routing, intra-chip communication, and parallel photonic neural networks. They correspond to large-scale vector-matrix products, which are the heart of neural network computation. Their implementation at the submicron scale through the use of TPP-DLW technology has already been tested [115], and the optical splitter is the basis of such interconnects. This splitter was thoroughly studied in [116]. It was designed to facilitate horizontal multiplexing into large arrays as well as vertical stacking into layers based on fractal geometries [115]. Thus, the splitter tested in [116] consisted of one input waveguide connected to four (2×2), nine (3×3), and sixteen (4×4) output waveguides, arranged in a square array with a lattice distance of D_0, a height of 52 μm, and a waveguide diameter of 1.2 μm, as represented in figures 3.24(a)–(c).

To enhance the adhesion of the splitters to the glass substrate during the TPP-DLW fabrication, a pedestal base covering the area of all the output waveguides was included at their terminals (see the printed results in figures 3.24(d)–(f)). The average writing power, P, influences the degree of polymerisation and consequently the writing voxel size [117]. On the other hand, the hatching distance, i.e. the distance between writing voxels, impacts the material's homogeneity and surface roughness. Control of these parameters is essential to reduce scattering losses and, as a result, reduce propagation losses. With this in mind, the authors tested two writing powers (i.e. $P = 10.4$ mW and $P = 11.2$ mW), chosen from experience to maintain stability and avoid burning, and two hatching distances ($h = 0.1$ μm and $h = 0.2$ μm). For the selected parameters, the voxel width was 0.6 μm and the voxel height 1.2 μm. The influence of the parameters on the surface quality can be seen in figure 3.25 [116].

Based on the schematic illustrations and SEM images in figure 3.25, it is clear that the hatching distance plays a crucial role in determining the surface quality of the splitters, where a smoother surface is achieved for smaller hatching distances. Additionally, it can be observed that for larger hatching distances, a periodic modulation identical to the hatching distance is observed at the surface of the splitter. Even when working at sub-nanometre scales, it is crucial to avoid these periodic structures, as they influence optical propagation.

To evaluate the waveguide splitters' performance in terms of transmission losses, they were characterised at 632 nm using a free-space apparatus to couple and collect the radiation passing in and out of the splitters' arms. The tests were conducted for

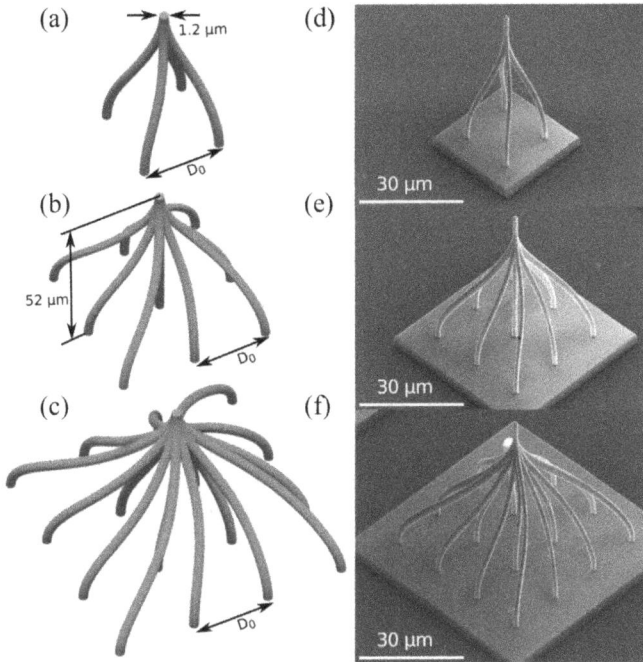

Figure 3.24. Designs for optical splitters with one input arm and (a) four (2 × 2), (b) nine (3 × 3), and (c) sixteen (4 × 4) output arms. SEM images of the TPP-DLW optical splitters with (d) four, (e) nine, and (f) sixteen output arms. Reprinted with permission from [116]. © 2020 Optical Society of America.

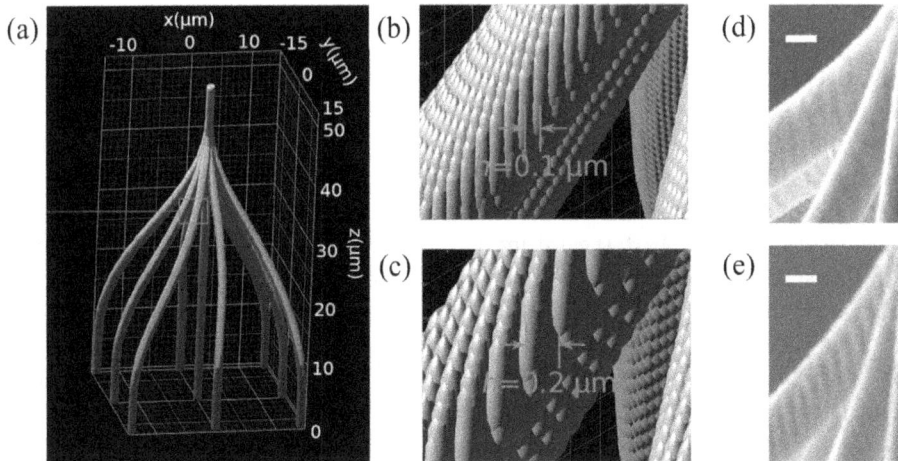

Figure 3.25. (a) Representation of a 1 × 9 optical splitter with $D_0 = 14$ μm. Panels (b) and (c) show magnified views of the red square region presented in (a), illustrating the effects of different hatching distances, $h = 0.1$ μm and $h = 0.2$ μm. Panels (d) and (e): SEM images of the waveguide surface for $h = 0.1$ μm and $h = 0.2$ μm, respectively. The scale bar represents 1 μm. Reprinted with permission from [116]. © 2020 Optical Society of America.

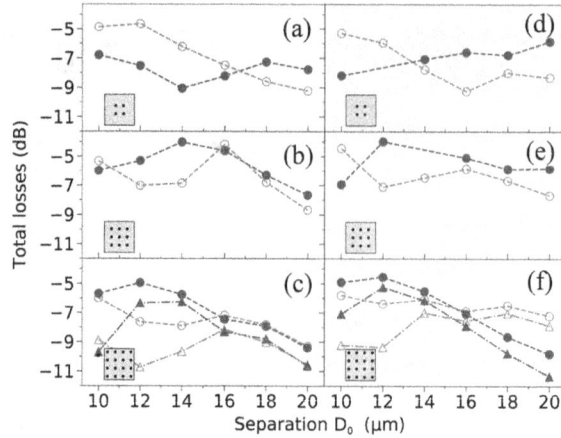

Figure 3.26. Optical losses measured between the input arm and varying numbers of output arms (i.e. four, nine, and sixteen from top to bottom) as a function of the distance between neighbouring output arms, making a total of three different splitter configurations. The results are shown for two writing powers, $P = 10.4$ mW and $P = 11.2$ mW in the left and right graphs, respectively, and for two hatching distances, $h = 0.2$ µm (shown in blue) and $h = 0.1$ µm (shown in red). In (c) and (f), the triangular points correspond to the central 2×2 sub-array of the 4×4 array. Adapted with permission from [116]. © 2020 Optical Society of America.

splitters with different values of D_0, h, and P, and the results are depicted in figure 3.26 [116].

The splitter transmission losses obtained from [116] and shown in figure 3.26 indicate that an increase in D_0 increases the losses, which are mostly associated with an increase in the number of modes at the splitter's bifurcation area, leading this region to work as a MM interferometer, as we will discuss later in this section for other splitter types. Regarding the writing power, there were no observable differences except for the 2×2 splitter configuration with $h = 0.1$ µm, which showed a slight decrease.

The splitters were also characterised using the output power distribution, which showed that the power was not equally distributed among the different output arms. In conclusion, the team stated that the use of SM waveguides and their proper projection to use MMI at the bifurcated splitter regions would be an important future prospect to enhance the performance of the produced splitters.

Recently, the additive manufacture of splitters capable of SM operation has also been introduced to explore 3D geometry. For this purpose, Gaso *et al* [118] designed a 1×4 splitter to support SM operation at 1550 nm. The proposed splitter was intended to be 3D printed using TPP by a commercial nano 3D printer that had horizontal and vertical resolutions of 200 and 600 nm, respectively. Due to the printer's build size limitation of 300 µm, the splitter length was also limited to this size. The splitter was composed of a 30 µm straight waveguide, intended to be connected to a standard SM telecom fibre. Additionally, the output branches were designed to be connected to similar standard fibres, which necessitated separating the branches to a distance equal to the fibre diameter, i.e., 127 µm. To keep the splitter length under the 300 μm limit, the branches followed arc cosine shapes with a maximum length of 270 μm, allowing them to attain a branch bending radius of 370.2 μm. According to the authors, a square-core waveguide was chosen to suppress polarisation- and wavelength-dependent losses. The input waveguide port was designed to match the core of the SM fibre; specifically, it had a cross-sectional area of 8×8 μm^2, followed by a tapered region resulting in a final waveguide cross-sectional area of 1.7×1.7 μm^2. The refractive indices of the core and cladding regions were 1.53 and 1.49, respectively, to simulate the materials used experimentally, namely IP-Dip negative photoresist and polydimethylsiloxane (PDMS) Sylgard 184. The splitter design, along with its dimensions, is schematically illustrated in figure 3.27(a). Despite the optimistic numerical results showing promise for low losses (1.3 dB) and good uniformity across the multiple output arms (0.005 dB), the authors realised that the dimensions needed to develop the proposed SM splitter were close to the technological limits of their tools; thus, they decided to relax the SM tolerances to a more realistic design of an MM splitter waveguide with cross-sectional core measurements of 4×4 μm^2, instead of 1.7×1.7 μm^2. However, even considering these more realistic dimensions, the fragility of the structure remained problematic; as a result, a supporting structure was added to provide mechanical stability to the printed waveguide. An SEM image of the waveguide

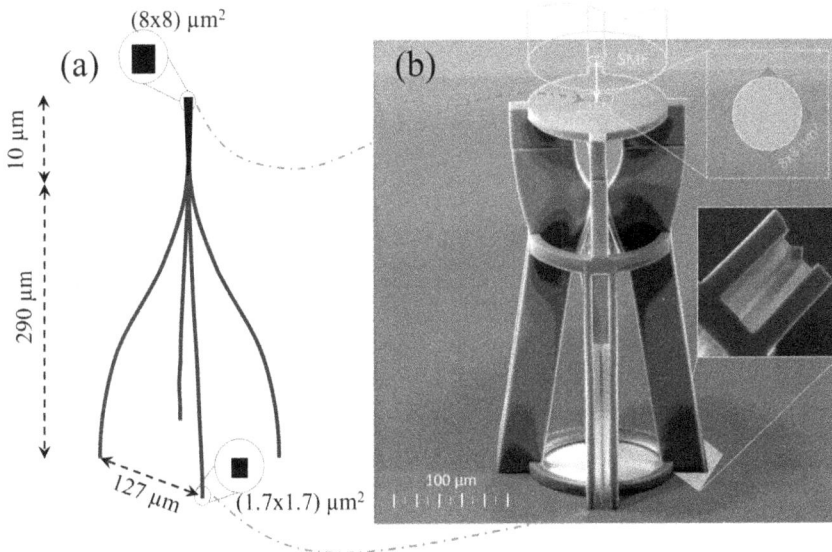

Figure 3.27. (a) Schematic of the 1 × 4 SM splitter designed to couple light from an SM fibre to four SM output fibres. The core had a square shape and was tapered down at the initial part of the input arm. The splitter's length was constrained by the printer's maximum volume, and the distance between the arms was set to allow four different fibres to touch each other. (b) SEM image of the beam splitter designed for MM behaviour (output arms with a cross-section of 4 × 4 μm) together with its supporting structure. In the inset image, it is possible to observe one of the square-shaped output waveguides supported by thin wall structures. © [2021] IEEE. Reprinted, with permission, from [118].

fabricated on top of a glass slide, following the removal of the unpolymerised resist, is shown in figure 3.27(b).

As can be seen in the inset of figure 3.27(b), the square output arm was supported by two thin walls positioned at the corners of the square-core waveguide. These supports were necessary to provide structural integrity to the waveguide. Furthermore, their square-corner positioning and thin dimensions were carefully chosen to minimise propagation losses. The splitter was further processed to achieve its final configuration. For this purpose, the tip of an SM fibre was aligned with the input arm of the splitter, and PDMS was then used to simulate the outer cladding region [118]. The PDMS was left to polymerise the printed part and fibre tip region. After carefully removing the glass lens, the authors observed good visible light distribution across the four output arms. Furthermore, near-field images obtained in the 1550 nm region revealed dominant Gaussian mode-field propagation. Finally, the authors estimated the waveguide loss to be about 3.6 dB, which was greater than the estimates given by the numerical results. They attributed this to coupling losses, mode-field mismatch between the SM fibre and the square-core waveguide, and finally, the taper region, which were not taken into account in the simulations. To conclude, despite their efforts to develop a waveguide splitter with SM behaviour, the authors were still only able to demonstrate the capability to create an MM splitter with relatively close dimensions to those of an SM waveguide. Although the team dedicated its efforts to reducing the waveguide core diameter to achieve SM behaviour, the core–cladding refractive index contrast was very high (0.04). Thus, this research could still be improved by choosing materials with a suitable core–cladding refractive index contrast, allowing for a reduction in the V-parameter and hence the ability to work with more relaxed waveguide dimensions.

3.5.3 Inversely tapered splitters

Due to technological limitations in developing 3D splitters with SM behaviour and low losses, and considering their benefits in terms of parallel scalability in photonic circuit applications, new splitting approaches have already been investigated. One of the paths followed was the integration of adiabatic coupling with AM. Adiabatic coupling is widely known in photonics [7] due to the efficient and broadband SM field transfer offered by one-to-N waveguides [119]. This type of coupling uses a sequence of tapered to inversely tapered waveguides. In this approach, the optical mode leaks from the core of the tapered structure through the cladding to the anti-tapered output waveguides [120]. Grabulosa et al [121] were among of the first to propose the use of 3D printing technology with adiabatic coupling. For this, they used a TPP printer to create the tapered and anti-tapered structure of the splitter. To achieve this, two tapering geometries were considered: conical and truncated rod. Representations of these geometries are shown on the left and right sides of figure 3.28(a), respectively, for a 1×2 splitter. Figure 3.28(b), on the other hand, shows the cross-sectional areas of $1 \times N$ splitters along their lengths.

The 3D-printed optical splitters were designed to facilitate adiabatic power transfers between the input arm and four output arms. For this, they inversely

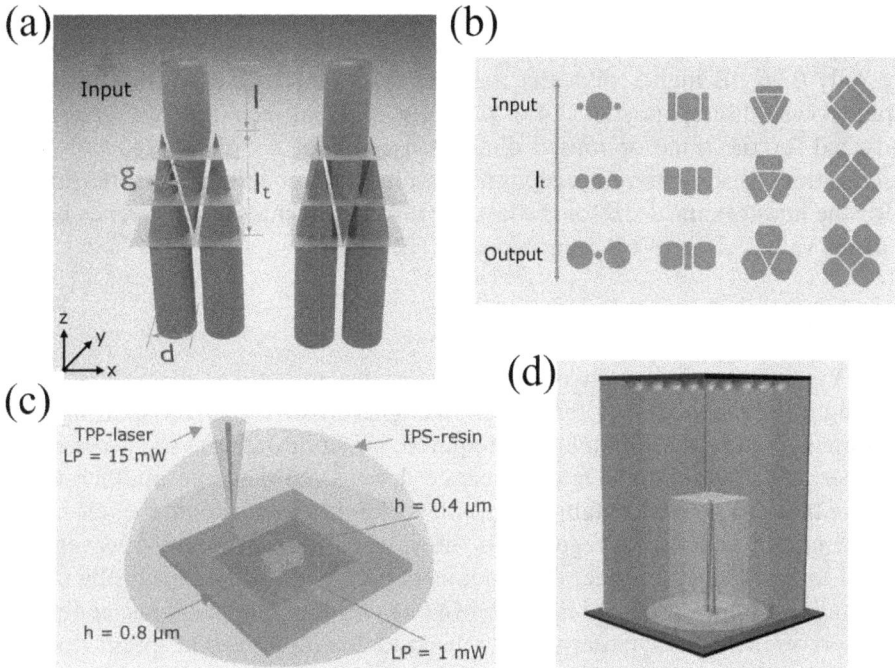

Figure 3.28. (a) Representative schematics of the tapered and anti-tapered structures of the input and output waveguides with conical (left) and truncated rod (right) tapering geometries. (b) Waveguide cross-sections along their whole lengths for 1 × 2, 1 × 2, 1 × 3, and 1 × 4 splitters (from left to right). (c) Representative schematic showing the fabrication of the splitter using TPP technology. (d) Full polymerisation of the structure in the cuboid created during the printing shown in (c). Reprinted with permission from [121]. © 2023 Optica Publishing Group.

tapered the input and output waveguides with equal taper rates and geometric symmetry to match the relevant effective modal indices. The arms were considered to have widths of 3.3 μm and to investigate the evanescent coupling, the authors tested different taper gap distances, g, specifically 0.4, 0.8, and 1.2 μm, and different taper lengths, l_t, of 100, 300, and 500 μm. The splitters were fabricated with a negative photoresist with a refractive index of 1.51. To obtain a good surface quality, the hatching distance was set to 0.4 μm, while the laser power was set to 15 mW. Furthermore, to guarantee a constant gap between the tapered regions and to ensure adhesion for each of the tapered arms, this gap region was slightly polymerised with a low TPP power (1 mW) [121]. Furthermore, the whole structure was kept inside a cube for later polymerisation. The fabrication process is schematised in figure 3.28(c), while the full polymerisation of the entire structure is shown in figure 3.28(d). The authors estimated a refractive index contrast between the core and cladding of approximately 5×10^{-3}.

The conclusions of the study reported in [121] were initially applied to a 1 × 2 splitter, indicating that truncated tapered profiles offer lower coupling losses due to the effective transfer area and better performance for lower taper waveguide separations and longer taper lengths. Finally, an optimised 1×2 splitter with $g = 0.4$ μm and $l_t = 500$ μm was tested, showing coupling losses of 0.32 dB, which were only 0.06 dB higher than the theoretically estimated value. In addition, an output power discrepancy of 3.4% was observed between the two arms. Tests conducted for the same optimised dimensions, but for 1×3, 1×4, and 1×16 configurations, also achieved exceptional results, with coupling losses reaching 0.4 dB for the first two and 1 dB for the last. Regarding the intensity difference between the arms, values close to 6% were reported [121].

3.5.4 Multimode interference splitters

While Y-splitters offer relatively simple designs and have no polarisation or wavelength dependence, they tend to show higher losses for larger splitting ratios. Additionally, when multiple splits are required, larger device sizes are necessary. On the other hand, multimode interferometer devices (MMIs) are another type of structure that can be used to fabricate splitters. These structures use the self-imaging phenomenon in an MM waveguide. In short, when light from a SM waveguide is coupled to an MM waveguide, the fundamental core mode couples to the multiple modes allowed to propagate in the MM waveguide region. This enables the formation of interference patterns that reproduce the input field at specific locations along the length of the MM waveguide. Through this phenomenon, one can split the light radiation from one arm into multiple arms, allowing for very uniform splitting ratios, compactness, and lower insertion losses than those of Y-splitters. This technology has been largely used in the PIC industry, allowing PIC designers to split light from an input arm to multiple output arms with different split ratios using a compact approach [122, 123]. The need to reduce the cost of manufacturing splitters and increase the miniaturisation of photonic devices has also led to the employment of 3D printing to manufacture MMI optical splitters.

Mizera *et al* [124] were among the first to report a 3D-printed MMI splitter through the use of TPP. The proposed splitter was designed to operate in the 1550 nm telecom region and to split light injected into one input arm from an SM optical fibre into four output arms. The splitter was numerically optimised to produce four equidistant maxima. For this, they considered the refractive index of the negative photoresist ($n = 1.53$) used for the TPP printing and air as the surrounding medium ($n = 1.00$). The optimised structure consisted of a parallelepiped with a cross-sectional area of (18×18) μm^2 and a length of 158 μm. The implemented 3D design of the structure is shown in the pink region of figure 3.29(a).

The proposed MMI splitter presented in [124] and shown in figure 3.29(a) was designed to be coupled to a standard telecom fibre, which led the authors to maximise the coupling efficiency between the fibre-guided mode and the square-core waveguide mode. For this purpose, they designed a tapered transition consisting of a 60 μm length and a cross-sectional transition from (10×10) μm^2 for the SM fibre side and (9×9) μm^2 for the MMI side. Due to the fragility of the whole structure (the MMI and taper region), a mechanical support structure was used. This was composed of thin bridges at each of the four corners of the MMI waveguide,

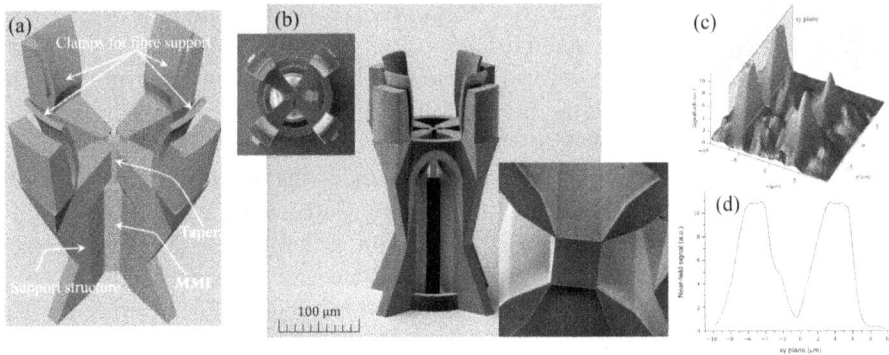

Figure 3.29. (a) Schematic of the proposed splitter composed of the MMI region (marked in pink) surrounded by a mechanical support structure (grey) used to provide mechanical robustness to the MMI and a clamping mechanism for the attachment of an SM fibre. (b) SEM image of the 3D MMI splitter, together with its supporting structure and the clamp. The inset at the top left shows the output side of the optical splitter, while the inset at the bottom left displays the MMI splitter output. Near-field mode distributions at the tip for the MMI for the (c) 3D and (d) 2D representations. Reproduced from [124]. CC BY 4.0.

connected to the outer support region (see the inset of figure 3.29(b)). This approach was mentioned earlier for the study reported in [118] and shown in figure 3.27(b), namely to ensure structural mechanical robustness while ensuring a low effect on the waveguide's guiding mechanisms. Additionally, to easily attach the MMI structure to the tip of an SM fibre, the mechanical structure was designed to include four protruding clamps capable of being anchored to the SM fibre tip (see figures 3.29(a) and (b)).

After TPP printing and removal of the unpolymerised resist in a developer bath, the structure shown in figure 3.29(b) was coupled to an SM fibre and later removed from the glass slide used as a substrate for its construction. The characterisation results were obtained through near-field mode observation at 1550 nm, revealing the presence of four maxima (see figures 3.29(c) and (d)), with little asymmetry caused by coupling issues but showing mode-field diameters of (6×6) μm^2, as anticipated from the simulations. Regarding the insertion losses, an overall loss of approximately 2.6 dB was measured for all branches, whereas the simulated value was close to 0.8 dB. These losses were attributed to the poor alignment of the SMF to the MMI structure. Despite the high losses, this was one of the first studies of its kind, and new MMI splitters have since followed. An example was reported by the same group in [125] for similar MMI structures consisting of 1×9 branches, showcasing the potential of these structures for vertical integration in photonic applications.

Splitters capable of dividing light propagating in three-dimensional space are also relevant in MCF technologies. Despite several record transmissions being achieved with these fibres [38] and promising prospects for shape sensing in fields such as medicine and robotics [126], the development of components for these fibres is scarce. Among those important components is the splitter, which allows the optical signal propagating along a single fibre core to be routed to the multiple cores available in the MCF. One of the first solutions explored for manufacturing MCF splitters was the splicing of standard telecommunication-grade single-core fibres (SCFs) with MCFs [127]. However, this method has significant limitations, including high losses and low repeatability. Another approach proposed was the use of fibre side polishing of an SCF to create evanescent field-based directional coupling between the core of the SCF and one of the cores of the MCF [39]. However, this technique has significant limitations since coupling to several cores of the MCF is almost impossible. In the field of 3D fabrication technologies, laser-based micro-machining of a glass substrate [40] has already been tested. However, only fan-in and fan-out devices have been developed for MCFs. This is due to the nature of the laser-based micromachining process, which is typically a lengthy and expensive process that produces components at large scales. Thus, this led to the introduction of AM into the production of MCF splitters. Baghdasaryan *et al* [128] introduced the use of the TPP-DLW technique to manufacture a proof of concept 1 × 4 MCF splitter able to operate in the C and L bands. The proposed MCF splitter was conceived based on tridimensional MMI technology and was inspired by previous work developed by other authors using SM fibres fused to non-circular MM fibres, allowing the distribution of the SM fibre input field into multiple equidistant positions in the MM fibre to match the core locations of the MCF [129, 130]. To develop the MCF splitter using TPP-DLW, Baghdasaryan *et al* [128] designed an MM waveguide with a star-like cross-section that distributed the input field to seven spot positions distributed along the cross-section of the star waveguide. However, to shorten the manufacturing time and to keep the proof of concept simple, the authors used an MM waveguide with a triangular cross-section instead of the star-shaped waveguide. This allowed the input field to be distributed to four outputs. The numerical results related to the location of the self-image formation for the triangular cross-section waveguide are shown in figure 3.30(a), while the schematic of the proposed device is shown in figures 3.30(b) and (c) and its estimated insertion losses are shown in figures 3.30(d) and (e) for the couplings between central–central waveguides and central–periphery waveguides, respectively.

The results of the insertion losses measured in the C + L bands revealed values between 0.5 and 1 dB with some polarisation dependence, as shown in figures 3.30(d) and (e), showing the feasibility of the design. The simulation results also revealed that the self-images formed in the cross-sectional area of the waveguide for the optimised length had a separation of 3 μm from each other, while the MCF core separation was 35 μm. Because of this, S-bend waveguides were designed to route the signal from each of the fibre cores to each of the MMI cross-sectional locations. Additionally, adiabatic tapers were also included to match the fundamental mode of the implemented fibres. The design of these structures is shown in red in figure 3.31 (a), while support structures are shown in grey. After a suitable design of the tapers and S-bends was completed, the splitter was additively manufactured through TPP-DLW on top of a glass slide. The fabricated 3D-printed splitters after treatment with a developer solution are shown in figures 3.31(c) and (d) [128].

After aligning the fibres to the newly formed waveguide splitter (see figures 3.31(e) and (f)), the authors achieved per-channel coupling losses of 3 dB. This was higher

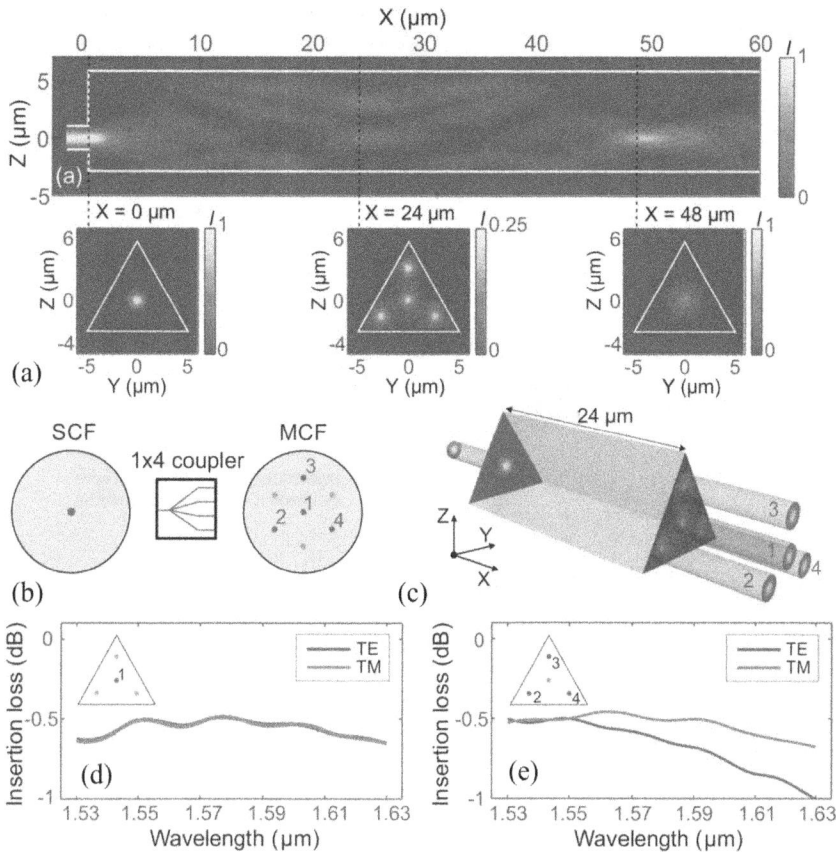

Figure 3.30. (a) Two-dimensional visualisation of the self-imaging phenomenon along the length of an MMI device with a triangular cross-section and a side wall of 10 μm. The bottom images show the spot field distributions in the cross-section of the MMI at different lengths. (b) Schematic illustration of the SM fibre to MCF splitter design using four target channels. (c) Proposed 1 × 4 MMI coupler with input and output waveguides 2 μm in diameter. (d) and (e) correspond to the numerically estimated insertion losses for coupling between the central input waveguide and the central and radial output waveguides, respectively [128] John Wiley & Sons. © 2024 The Author(s). Laser & Photonics Reviews published by Wiley-VCH GmbH.

than the numerically estimated value (1.2 dB) [128]. Furthermore, polarisation-dependent losses ranging from 1 to 3 dB were also observed. Despite the excessive losses, this study showed impressively complex fabrication details. However, research can still be conducted to further reduce the achieved insertion losses. The results described for the different splitters demonstrated in the literature indicate that their manufacture is already possible using various AM technologies. Despite this, the solutions reported so far are still in their early stages and reveal high insertion losses. As technology advances, we can expect these solutions to become crucial technological assets for next-generation optical routing, intra-chip optical communications, free-space and MCF applications, and parallel photonic neural networks, among others.

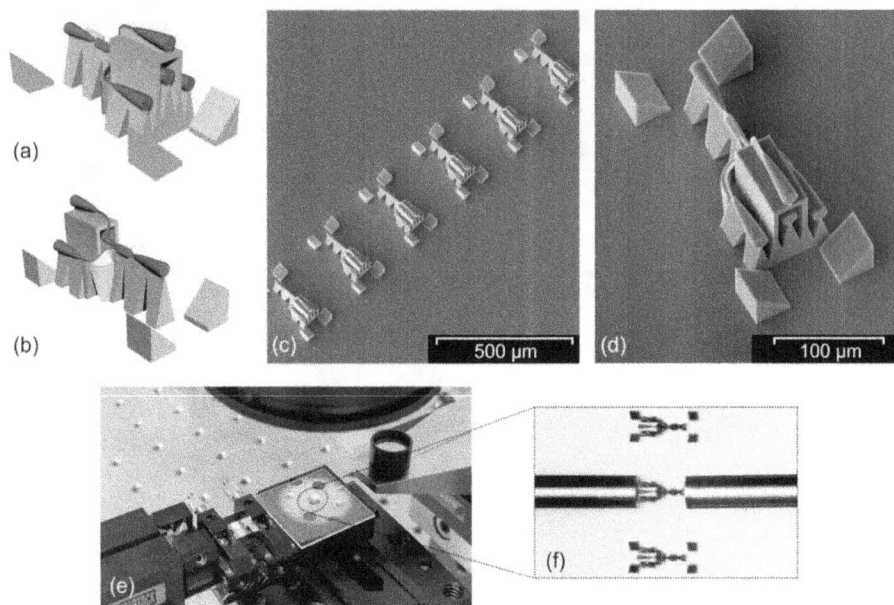

Figure 3.31. (a) and (b) Representative schematics of the splitter (in red) and the supporting structures from front and rear views. (c) SEM image of an array of splitters. (d) SEM image of a splitter. (e) Setup used to characterise the MCF splitter, with an MCF and an SCF fixed at the extremities of the splitter by fibre holders. (f) Microscope image of the splitter together with two fibres ready for characterisation [128] John Wiley & Sons. © 2024 The Author(s). Laser & Photonics Reviews published by Wiley-VCH GmbH.

References

[1] Kachris C and Tomkos I 2013 Power consumption evaluation of all-optical data center networks *Cluster Comput.* **16** 611–23

[2] Van Heddeghem W, Idzikowski F, Vereecken W, Colle D, Pickavet M and Demeester P 2012 Power consumption modeling in optical multilayer networks *Photonic Netw. Commun.* **24** 86–102

[3] Kita D M *et al* 2018 High-performance and scalable on-chip digital Fourier transform spectroscopy *Nat. Commun.* **9** 4405

[4] Passaro V M N, de Tullio C, Troia B, Notte M L, Giannoccaro G and Leonardis F D 2012 Recent advances in integrated photonic sensors *Sensors* **12** 15558–98

[5] Wang J and Long Y 2018 On-chip silicon photonic signaling and processing: a review *Sci. Bull. (Beijing)* **63** 1267–310

[6] Cheben P, Halir R, Schmid J H, Atwater H A and Smith D R 2018 Subwavelength integrated photonics *Nature* **560** 565–72

[7] Marchetti R, Lacava C, Carroll L, Gradkowski K and Minzioni P 2019 Coupling strategies for silicon photonics integrated chips *Photonics Res.* **7** 201–39

[8] Ranno L *et al* 2022 Integrated photonics packaging: challenges and opportunities *ACS Photonics* **9** 3467–85

[9] Lee J S *et al* 2016 Meeting the electrical, optical, and thermal design challenges of photonic-packaging *IEEE J. Sel. Top. Quantum Electron.* **22** 409–17

[10] Pavarelli N *et al* 2015 Optical and electronic packaging processes for silicon photonic systems *J. Lightwave Technol.* **33** 991–7

[11] Liu S F, Hou Z W, Lin L, Li Z and Sun H B 2023 3D laser nanoprinting of functional materials *Adv. Funct. Mater.* **33** 2211280

[12] Gonzalez-Hernandez D, Varapnickas S, Bertoncini A, Liberale C and Malinauskas M 2023 Micro-optics 3D printed via multi-photon laser lithography *Adv. Opt. Mater.* **11** 2201701

[13] Hatori N *et al* 2014 A hybrid integrated light source on a silicon platform using a trident spot-size converter *J. Lightwave Technol.* **32** 1329–36

[14] Snyder B, Corbett B and Obrien P 2013 Hybrid integration of the wavelength-tunable laser with a silicon photonic integrated circuit *J. Lightwave Technol.* **31** 3934–42

[15] Van Der Tol J J G M, Oei Y S, Khalique U, Ntzel R and Smit M K 2010 InP-based photonic circuits: comparison of monolithic integration techniques *Prog. Quantum Electron.* **34** 135–72

[16] Lindenmann N *et al* 2012 Photonic wire bonding: a novel concept for chipscale interconnects *Opt. Express* **20** 17667–77

[17] Vermeulen D *et al* 2010 High-efficiency fiber-to-chip grating couplers realized using an advanced CMOS-compatible silicon-on-insulator platform *Opt. Express* **18** 18278–83

[18] Mcnab S J, Moll N and Vlasov Y A 2003 Ultra-low loss photonic integrated circuit with membrane-type photonic crystal waveguides *Opt. Express* **11** 2927–39

[19] Abbasi A *et al* 2017 43 Gb/s NRZ-OOK direct modulation of a heterogeneously integrated InP/Si DFB laser *J. Lightwave Technol.* **35** 1235–40

[20] Luo X *et al* 2015 High-throughput multiple dies-to-wafer bonding technology and III/V-on-Si hybrid lasers for heterogeneous integration of optoelectronic integrated circuits *Front. Mater.* **2** 1–28

[21] Lu H *et al* 2016 Flip-chip integration of tilted VCSELs onto a silicon photonic integrated circuit *Opt. Express* **24** 16258–66

[22] Buffolo M *et al* 2017 Degradation mechanisms of heterogeneous III–V/silicon 1.55-μm DBR laser diodes *IEEE J. Quantum Electron.* **53** pp 1–8

[23] De Groote A *et al* 2016 Transfer-printing-based integration of single-mode waveguide-coupled III-V-on-silicon broadband light emitters *Opt. Express* **24** 13754–62

[24] Lin S *et al* 2016 Efficient, tunable flip-chip-integrated III–V/Si hybrid external-cavity laser array *Opt. Express* **24** 21454–62

[25] Billah M R *et al* 2018 Hybrid integration of silicon photonics circuits and InP lasers by photonic wire bonding *Optica* **5** 876–83

[26] Lindenmann N *et al* 2015 Connecting silicon photonic circuits to multicore fibers by photonic wire bonding *J. Lightwave Technol.* **33** 755–60

[27] Blaicher M *et al* 2020 Hybrid multi-chip assembly of optical communication engines by *in situ* 3D nano-lithography *Light: Sci. Appl.* **9** 71

[28] Sun J, Timurdogan E, Yaacobi A, Hosseini E S and Watts M R 2013 Large-scale nanophotonic phased array *Nature* **493** 195–9

[29] Seok T J, Quack N, Han S, Muller R S and Wu M C 2016 Large-scale broadband digital silicon photonic switches with vertical adiabatic couplers *Optica* **3** 64

[30] Ma Y *et al* 2013 Ultralow loss single layer submicron silicon waveguide crossing for SOI optical interconnect *Opt. Express* **21** 29374–82

[31] Zhang Y, Hosseini A, Xu X, Kwong D and Chen R T 2013 Ultralow-loss silicon waveguide crossing using Bloch modes in index-engineered cascaded multimode-interference couplers *Opt. Lett.* **38** 3608–11

[32] Itoh K *et al* 2016 Crystalline/amorphous Si integrated optical couplers for 2D/3D interconnection *IEEE J. Sel. Top. Quantum Electron.* **22** 255–63

[33] Shang K, Pathak S, Guan B, Liu G and Yoo S J B 2015 Low-loss compact multilayer silicon nitride platform for 3D photonic integrated circuits *Opt. Express* **23** 21334–42

[34] Sacher W D, Huang Y, Lo G Q and Poon J K S 2015 Multilayer silicon nitride-on-silicon integrated photonic platforms and devices *J. Lightwave Technol.* **33** 901–10

[35] Sacher W D *et al* 2017 Tri-layer silicon nitride-on-silicon photonic platform for ultra-low-loss crossings and interlayer transitions *Opt. Express* **25** 30862

[36] Suzuki K *et al* 2017 Ultralow-crosstalk and broadband multi-port optical switch using SiN/Si double-layer platform *Opto-Electronics and Communications Conf. (OECC) and Photonics Global Conf. (PGC)* 1–2

[37] Nesic A *et al* 2019 Photonic-integrated circuits with non-planar topologies realized by 3D-printed waveguide overpasses *Opt. Express* **27** 17402–25

[38] Puttnam B J, Rademacher G and Luís R S 2021 Space-division multiplexing for optical fiber communications *Optica* **8** 1186–203

[39] Zhang H *et al* 2017 A tuneable multi-core to single mode fiber coupler *IEEE Photonics Technol. Lett.* **29** 591–4

[40] Thomson R R *et al* 2007 Ultrafast-laser inscription of a three dimensional fan-out device for multicore fiber coupling applications *Opt. Lett.* **15** 11691–7

[41] Doerr C R and Taunay T F 2011 Silicon photonics core-, wavelength-, and polarization-diversity receiver *IEEE Photonics Technol. Lett.* **23** 597–9

[42] Kopp C *et al* 2011 Silicon photonic circuits: on-CMOS integration, fiber optical coupling, and packaging *IEEE J. Sel. Top. Quantum Electron.* **17** 498–509

[43] Taillaert D *et al* 2002 An out-of-plane grating coupler for efficient butt-coupling between compact planar waveguides and single-mode fibers *IEEE J. Quantum Electron.* **38** 949–55

[44] Yeh S M, Huang S Y and Cheng W H 2005 A new scheme of conical-wedge-shaped fiber endface for coupling between high-power laser diodes and single-mode fibers *J. Lightwave Technol.* **23** 1781–6

[45] Song J H, Fernando H N J, Roycroft B, Corbett B and Peters F H 2009 Practical design of lensed fibers for semiconductor laser packaging using laser welding technique *J. Lightwave Technol.* **27** 1533–9

[46] Tian Z-N *et al* 2013 Beam shaping of edge-emitting diode lasers using a single double-axial hyperboloidal micro-lens *Opt. Lett.* **38** 5414–7

[47] Williams H E, Freppon D J, Kuebler S M, Rumpf R C and Melino M A 2011 Fabrication of three-dimensional micro-photonic structures on the tip of optical fibers using SU-8 *Opt. Express* **19** 22910–22

[48] Dietrich P-I *et al* 2016 Lenses for low-loss chip-to-fiber and fiber-to-fiber coupling fabricated by 3D direct-write lithography *Conf. on Lasers and Electro-Optics.*

[49] Schneider S, Lauermann M, Dietrich P-I, Weimann C, Freude W and Koos C 2016 Optical coherence tomography system mass-producible on a silicon photonic chip *Opt. Express* **24** 1573–86

[50] Dietrich P I *et al* 2018 In situ 3D nanoprinting of free-form coupling elements for hybrid photonic integration *Nat. Photonics* **12** 241–7

[51] Suzuki T *et al* 2016 Cost-effective optical sub-assembly using lens-integrated surface-emitting laser *J. Lightwave Technol.* **34** 358–64

[52] Moehrle M *et al* 2010 Ultra-low threshold 1490 nm surface-emitting BH-DFB laser diode with integrated monitor photodiode *22nd Int. Conf. on Indium Phosphide and Related Materials (IPRM)* 1–4

[53] Kuzyk M G, Paek U C and Dirk C W 1991 Guest–host polymer fibers for nonlinear optics *Appl. Phys. Lett.* **59** 902–4

[54] Peng G D, Chu P L, Xiong Z, Whitbread T W and Chaplin R P 1996 Dye-doped step-index polymer optical fiber for broadband optical amplification *J. Lightwave Technol.* **14** 2215–23

[55] Bonefacino J *et al* 2018 Ultra-fast polymer optical fibre Bragg grating inscription for medical devices *Light: Sci. Appl.* **7** 17161

[56] Woyessa G, Fasano A, Stefani A, Markos C, Rasmussen H K and Bang O 2016 Single mode step-index polymer optical fiber for humidity insensitive high temperature fiber Bragg grating sensors *Opt. Express* **24** 1253–60

[57] Luo Y, Canning J, Zhang J and Peng G D 2020 Toward optical fibre fabrication using 3D printing technology *Opt. Fiber Technol.* **58** 102299

[58] Luo Y, Chu Y, Zhang J, Wen J and Peng G D 2025 3D printing-based photonic waveguides, fibers, and applications *Appl. Phys. Rev.* **12**

[59] Willis K D D, Brockmeyer E, Hudson S E and Poupyrev I 2012 Printed optics: 3D printing of embedded optical elements for interactive devices *Proc. of the 25th Annual ACM Symp. on User Interface Software and Technology* (New York) pp. 589–98

[60] Pereira T, Rusinkiewicz S and Matusik W 2014 Computational light routing: 3D printed fiber optics for sensing and display *ACM Trans. Graph.* **33** 24

[61] Canning J, Hossain M A, Han C, Chartier L, Cook K and Athanaze T 2016 Drawing optical fibers from three-dimensional printers *Opt. Lett.* **41** 5551

[62] Lorang D J, Tanaka D, Spadaccini C M, Rose K A, Cherepy N J and Lewis J A 2011 Photocurable liquid core-fugitive shell printing of optical waveguides *Adv. Mater.* **23** 5055–8

[63] Cook K *et al* 2016 Step-index optical fibre drawn from 3D printed preforms *Opt. Lett.* **41** 4554–7

[64] Gozzard D R, Craine R, Hickey D, Martin A, Shen W and Sones B 2022 Optical couplers and step-index fibers fabricated using FDM 3D printers *Opt. Lett.* **47** 5124

[65] Takahashi H, Punpongsanon P and Kim J 2020 Programmable filament: printed filaments for multi-material 3D printing *UIST 2020—Proc. of the 33rd Annual ACM Symp. on User Interface Software and Technology* (New York: Association for Computing Machinery, Inc.) 1209–21

[66] Zhao Q *et al* 2017 Optical fibers with special shaped cores drawn from 3D printed preforms *Optik (Stuttg)* **133** 60–5

[67] Birks T A, Knight J C and Russell P S 1997 Endlessly single-mode photonic crystal fiber *Opt. Lett.* **22** 961–3

[68] Oliveira R, Bilro L and Nogueira R N 2019 Principles of polymer optical fibers *Plastic Optical Fiber Sensors* 1st edn (Boca Raton, FL: CRC Press) 21–65

[69] Cook K *et al* 2015 Air-structured optical fiber drawn from a 3D-printed preform *Opt. Lett.* **40** 3966

[70] Zubel M G, Fasano A, Woyessa G, Sugden K, Rasmussen H K and Bang O 2016 3D-printed PMMA preform for hollow-core POF drawing *25th Int. Conf. on Plastic Optical Fibers* 295–300

[71] Marques T H R, Lima B M, Osório J H, Silva L E and Cordeiro C M B 2017 3D printed microstructured optical fibers *SBMO/IEEE MTT-S Int. Microwave and Optoelectronics Conf. (IMOC)* (Brazil: Aguas de Lindoia)

[72] Talataisong W *et al* 2018 Novel method for manufacturing optical fiber: extrusion and drawing of microstructured polymer optical fibers from a 3D printer *Opt. Express* **26** 32007

[73] Talataisong W *et al* 2018 Mid-IR hollow-core microstructured fiber drawn from a 3D printed PETG preform *Sci. Rep.* **8** 8113

[74] Zubel M G *et al* 2020 Bragg gratings inscribed in solid-core microstructured single-mode polymer optical fiber drawn from a 3D-printed polycarbonate preform *IEEE Sens. J.* **20** 12744–57

[75] Rahman M, Dilsiz C and Ordu M 2022 3D printed hollow-core polymer optical fiber with six-pointed star cladding for the light guidance in the near-IR regime *Int. Conf. on Information Science and Communications Technologies, ICISCT 2022 (Tashkent, Uzbekistan)* **2022** *(Piscataway, NJ: IEEE)*

[76] Kotz F *et al* 2017 Three-dimensional printing of transparent fused silica glass *Nature* **544** 337–9

[77] Kotz F *et al* 2019 Fabrication of arbitrary three-dimensional suspended hollow microstructures in transparent fused silica glass *Nat. Commun.* **10** 1439

[78] Nguyen D T *et al* 2017 3D-printed transparent glass *Adv. Mater.* **29** 1701181

[79] Destino J F *et al* 2018 3D printed optical quality silica and silica–titania glasses from sol–gel feedstocks *Adv. Mater. Technol.* **3** 1700323

[80] Chu Y *et al* 2019 Silica optical fiber drawn from 3D printed preforms *Opt. Lett.* **44** 5358

[81] Camacho Rosales A L, Núñez-Velázquez M and Sahu J K 2020 3D printed Er-doped silica fibre by direct ink writing *EPJ Web. Conf.* **243** 20002

[82] Camacho-Rosales A, Núñez-Velázquez M, Zhao X, Yang S and Sahu J K 2019 Development of 3-D printed silica preforms *2019 Conf. on Lasers and Electro-Optics Europe and European Quantum Electronics Conf.* (Munich: Optica Publishing Group) ce_7

[83] Camacho Rosales A L, Núñez Velázquez M A, Zhao X and Sahu J K 2020 Optical fibers fabricated from 3D printed silica preforms *Laser 3D Manufacturing VII* ed H Helvajian, B Gu and H Chen (Bellingham, WA: SPIE) p. 29

[84] Hänzi P *et al* 2025 Laser-assisted rapid prototyping of silica optical fibers functionalized with nanodiamonds and multiple active rare earth dopants *Opt. Mater. Express* **15** 949/966

[85] Maniewski P, Wörmann T J, Pasiskevicius V, Holmes C, Gates J C and Laurell F 2024 Advances in laser-based manufacturing techniques for specialty optical fiber *J. Am. Ceram. Soc.* **107** 5143–58

[86] Carcreff J *et al* 2021 Investigation on chalcogenide glass additive manufacturing for shaping mid-infrared optical components and microstructured optical fibers *Crystals (Basel)* **11** 228

[87] Gołębiewski P *et al* 2024 3D soft glass printing of preforms for microstructured optical fibers *Addit. Manuf.* **79** 103899

[88] Zheng B, Yang J, Qi F, Wang J, Zhang X and Wang P 2021 Fabrication of Yb-doped silica micro-structured optical fibers from UV-curable nano-composites and their application in temperature sensing *J. Non. Cryst. Solids.* **573** 121129

[89] Luo Y *et al* 2024 All solid photonic crystal fiber enabled by 3D printing fiber technology for sensing of multiple parameters *Adv. Sens. Res.* **3** 2300205

[90] Canning J, Chu Y, Luo Y, Peng G D and Zhang J 2022 Challenges in the additive manufacture of single and multi-core optical fibres *J. Phys. Conf. Ser.* **2172** 012008

[91] Luo Y *et al* 2024 3D printing specialty multi-function twin core Bi/Er co-doped silica optical fibres for ultra-broadband polarized near infrared emission and sensing applications *Opt. Laser Technol.* **168** 109817

[92] Chu Y *et al* 2022 Additive manufacturing fiber preforms for structured silica fibers with bismuth and erbium dopants *Light Adv. Manuf.* **3** 358–64

[93] Wang J *et al* 2024 Fabrication of Hi-Bi multi-core silica optical fibre preforms from dual-curing resins incorporating nano composites *J. Lightwave Technol.* **42** 6547–54

[94] Cariñe J *et al* 2020 Multi-core fiber integrated multi-port beam splitters for quantum information processing *Optica* **7** 542–50

[95] Miller D A B 2015 Perfect optics with imperfect components *Optica* **2** 747–50

[96] Nourshargh N, Starr E M and Ong T M 1989 Integrated optic 1×4 splitter in SiO_2/GeO_2 *Electron. Lett.* **25** 981–2

[97] Soldano L B, Veerman F B, Smit M K, Verbeek B H, Dubost A H and Pennings E C M 1992 Planar monomode optical couplers based on multimode interference effects *J. Lightwave Technol.* **10** 1843–50

[98] Mizuno H, Sugihara O, Jordan S, Okamoto N, Ohama M and Kaino T 2006 Replicated polymeric optical waveguide devices with large core connectable to plastic optical fiber using thermo-plastic and thermo-curable resins *J. Lightwave Technol.* **24** 919–26

[99] Rezem M, Gunther A, Roth B, Reithmeier E and Rahlves M 2017 Low-cost fabrication of all-polymer components for integrated photonics *J. Lightwave Technol.* **35** 299–308

[100] Die Mount GmbH https://diemount.com (accessed 06 March 2024).

[101] Park H-J, Lim K-S and Kang H S 2011 Low-cost 1×2 plastic optical beam splitter using a V-type angle polymer waveguide for the automotive network *Opt. Eng.* **50** 075002

[102] Yamashita T, Kawasaki A, Kagami M, Yasuda T and Goto H 2010 Polymeric multi/demultiplexers using light-induced self-written waveguides for cost-effective optical interconnection *2010 IEEE CPMT Symp.* (Piscataway, NJ: IEEE) 1–4

[103] Zhou Z and Duan X 2006 Integrated waveguide splitter fabricated by Cs^+–Na^+ ion-exchange *Opt. Commun.* **266** 129–31

[104] Bamiedakis N, Beals IV J, Penty R V, White I H, DeGroot J V and Clapp T V 2009 Cost-effective multimode polymer waveguides for high-speed on-board optical interconnects *IEEE J. Quantum Electron.* **45** 415–24

[105] Klotzbücher T, Braune T, Dadic D, Sprzagala M and Koch A 2003 Fabrication of optical 1x2 POF couplers using the Laser-LIGA technique *Laser Micromachining for Optoelectronic Device Fabrication* (Bellingham, WA: SPIE) pp. 121–32

[106] Takezawa Y, Akasaka S-I, Ohara S, Ishibashi T, Asano H and Taketani N 1994 Low excess losses in a Y-branching plastic optical waveguide formed through injection molding *Appl. Opt.* **33** 2307–12

[107] Annuar A, Shaari S, Kamil M and Rahman A 2011 Acrylic and metal based Y-branch plastic optical fiber splitter with optical NOA63 polymer waveguide taper region *Opt. Rev.* **18** 80–5

[108] Prajzler V and Zázvorka J 2019 Polymer large core optical splitter 1× 2 Y for high-temperature operation *Opt. Quantum Electron.* **51** 216

[109] Calvert P 2001 Inkjet printing for materials and devices *Chem. Mater.* **13** 3299–305

[110] Cox W R, Chen T and Hayes D J 2001 Micro-optics fabrication by ink-jet printing *Opt. Photonics News* **12** 32–5

[111] Hayes D J, Co W R and Grove M E 1998 Micro-jet printing of polymers and solder for electronics manufacturing *J. Electron. Manuf.* **8** 209–16

[112] Li S, Lin Q, Chen L and Wu X 2011 Design and fabrication of polymeric multimode power splitter with secondary asymmetric branches *Chin. Opt. Lett.* **9** 081302

[113] Prajzler V and Zavřel J 2021 Large core optical elastomer splitter fabricated by using 3D printing pattern *Opt. Quantum Electron.* **53** 337

[114] Oliveira R, Nogueira R and Bilro L 2021 Do-it-yourself three-dimensional large core multimode fiber splitters through consumer-grade 3D printer *Opt. Mater. Express* **12** 593–605

[115] Moughames J *et al* 2020 Three-dimensional waveguide interconnects for scalable integration of photonic neural networks *Optica* **7** 640

[116] Moughames J, Porte X, Larger L, Jacquot M, Kadic M and Brunner D 2020 3D printed multimode-splitters for photonic interconnects *Opt. Mater. Express* **10** 2952

[117] Jiang L J *et al* 2014 Two-photon polymerization: investigation of chemical and mechanical properties of resins using Raman microspectroscopy *Opt. Lett.* **39** 3034–7

[118] Gaso P *et al* 2021 3D polymer based 1×4 beam splitter *J. Lightwave Technol.* **39** 154–61

[119] Dewanjee A, Caspers J N, Aitchison J S and Mojahedi M 2016 Demonstration of a compact bilayer inverse taper coupler for Si-photonics with enhanced polarization insensitivity *Opt. Express* **24** 28194–203

[120] Snyder A W 1970 Coupling of modes on a tapered dielectric cylinder *IEEE Trans. Microw. Theory. Tech.* **18** 383–92

[121] Grabulosa A, Porte X, Jung E, Moughames J, Kadic M and Brunner D 2023 (3 + 1)D printed adiabatic 1-to-M broadband couplers and fractal splitter networks *Opt. Express* **31** 20256–64

[122] Deng Q, Liu L, Li X and Zhou Z 2014 Arbitrary-ratio 1× 2 power splitter based on asymmetric multimode interference *Opt. Lett.* **39** 5590–3

[123] Samoi E, Benezra Y and Malka D 2020 An ultracompact 3×1 MMI power-combiner based on Si slot-waveguide structures *Photonics Nanostruct.* **39** 100780

[124] Mizera T, Gaso P, Pudis D, Ziman M, Kuzma A and Goraus M 2022 3D Polymer-based 1× 4 MMI splitter *Nanomaterials* **12** 1749

[125] Ziman M, Feiler M, Mizera T, Kuzma A, Pudis D and Uherek F 2022 Design of a power splitter based on a 3D MMI coupler at the fibre-tip *Electronics (Switzerland)* **11** 2815

[126] Jäckle S, Eixmann T, Schulz-Hildebrandt H, Hüttmann G and Pätz T 2019 Fiber optical shape sensing of flexible instruments for endovascular navigation *Int. J. Comput. Assist. Radiol. Surg.* **14** 2137–45

[127] Pytel A *et al* 2017 Optical power 1 × 7 splitter based on multicore fiber technology *Opt. Fiber Technol.* **37** 1–5

[128] Baghdasaryan T, Vanmol K, Thienpont H, Berghmans F and Van Erps J 2024 Ultracompact 3D splitter for single-core to multi-core optical fiber connections fabricated through direct laser writing in polymer *Laser Photon. Rev.* 2400089

[129] Zhang Z *et al* 2018 Fiber-based three-dimensional multi-mode interference device as efficient power divider and vector curvature sensor *J. Opt.* **20** 035701

[130] Zhang Z *et al* 2018 Single-mode to 61 spots divider with multimode interference in hexagonal core fiber *IEEE Photonics Technol. Lett.* **30** 1337–40

IOP Publishing

Additive Manufacturing in Optics and Photonics
Fabrication and applications
Ricardo Oliveira and Nuno Valente

Chapter 4

Micro- and nanophotonic 3D printing: devices and applications

Additive manufacturing (AM) has been widely explored for the manufacture of components. Its advantages, such as rapid prototyping, low material waste, low production costs, and high design freedom, are some of the characteristics that make this technology extremely desirable. As a result, several 3D printing techniques have already evolved to support different fields of engineering. This allowed the capability to print components that were previously only possible to imagine in a computer simulation. The range of printing materials is another significant advancement in this field. Examples of such materials include polymers, metals, and ceramics. When it comes to printing size, the word 'additive' indicates that, theoretically, there should be no restriction on the size of a component. However, bigger does not always mean better. Optical technologies are one example where 3D printing needs to work at a small scale. These technologies require the manipulation and control of light. Therefore, their resolutions should be smaller than the wavelength they are intended to work with.

When the objective is to print optical components with resolutions at the micrometre or nanometre scale, i.e. scales comparable to the wavelength of light, it is necessary to employ high-resolution 3D printing. This is a necessity in fields other than optics, as other technological fields can also benefit from these small-scale features [1–3]. To achieve high-resolution 3D printing, it is generally necessary to confine the light beam energy to a specific three-dimensional space, allowing it to induce a reaction in a particular volume of material. A straightforward solution is to illuminate a photopolymer with a single light beam, leading to photon absorption by a molecule of the photopolymer and, consequently, polymerisation. This method is commonly referred to as one-photon absorption (OPA) [4–6]. However, this method has some limitations, such as the fact that the material in the optical path of the beam is fully polymerised, leading to challenges related to penetration depth and also poor resolution. Therefore, more advanced methods should be employed.

doi:10.1088/978-0-7503-6428-7ch4   4-1

An example of this is two-photon absorption (TPA). In this method, the polymerisation of a particular region in three-dimensional space, commonly known as a voxel, requires the simultaneous absorption of two photons. This allows for spot size resolutions as low as 100 nm [7]. This is only possible when the light intensity reaches a certain threshold. This is more likely to occur in the region where the light intensity is highest, namely the focal point of the laser beam. Thus, this tight localisation leads to resolutions higher than those achieved by the OPA process [8–10].

The TPA technique was first described in 1931 by Goeppert Mayer [11] and later experimentally demonstrated in 1961 [12]. Over the years, advancements in the TPA process, such as the development of femtosecond (fs) laser technology, led to the development of two-photon polymerisation lithography (TPL) technology in the 1990s, also known as two-photon polymerisation (TPP). This technology is well disseminated and is widely used to 3D print structures at the nanometre scale and with very low surface roughness (<10 nm). Because of this, several companies, such as Nanoscribe, UpNano, Multiphoton Optics, Vanguard Photonics, Microlight3D, and Femtika, have marketed different versions of their TPP products, marking a new transitional era in the field of nanotechnology.

4.1 TPP printer configuration and materials

In simple terms, a TPP 3D printing system is composed of a collimated fs laser beam, an attenuator, a beam expander, a shutter, a pair of galvanometers for two-dimensional beam scanning (or two piezo linear systems in some 3D printing systems), a dichroic mirror for visualisation, and an objective lens to focus the laser beam [13]. The materials used in TPP printing, commonly referred to as photopolymers, consist of two components: the monomer/oligomer mixture (or at least one of these) and the photoinitiator. The first acts as the crosslinker, forming the basic structure of the 3D structure, while the second is a chemical compound that absorbs light and initiates a chemical reaction, leading to the formation of active species and, consequently, the polymerisation process [14, 15]. It is important to note that the material must be transparent to the wavelength of the fs laser radiation, allowing it to propagate and cause polymerisation only in the focal region, where the light intensity exceeds the polymerisation threshold but remains below the ablation threshold.

To date, most of the materials used in TPP have been designed for conventional photolithographic applications, with examples of both negative [16, 17] and positive photoresists [18, 19]. Concerning negative photoresists, which are the most frequently used, two-photon exposure leads to polymerisation and hardening in the focal region, where multiphoton absorption occurs. The remaining unpolymerised liquid resist can then be easily washed away with a developer solution. The most common commercial example is the epoxy-based photoresist SU-8. However, SU-8 is not optimised for multiphoton absorption and lacks the voxel-level precision needed for fine photonic features. Resists used for this purpose include acrylic-based resists such as the IP-Series (e.g. IP-L, IP-Dip, IP-S) from Nanoscribe GmbH or hybrid organic–inorganic polymers developed by Micro Resist Technology GmbH, such as OrmoComp® from the ORMOCER family. Regarding positive photoresists, a resist layer is soft-baked at low temperatures (60 °C–80 °C) on a substrate, and then specific regions are exposed to radiation, leading to a chemical change where the photoactive compound within the resist is converted into a more soluble form, which can later be dissolved and removed in the washing step.

TPP materials can be grouped into organic and hybrid materials. Regarding the first group, there are three main materials that fit into this category: acrylates, hydrogels, and epoxies. Acrylates were the first materials to be used in the TPP process [20]. The reasons relate to their low cost, wide availability, ease of processing, and high transparency in the visible and infrared wavelengths. Several studies have reported the use of this material, allowing for the exploration of key features such as accurate optical structures with low roughness and low shrinkage [21–23]. Hydrogels are organic photopolymers that offer exciting possibilities in biomedical and tissue engineering applications. Their biocompatibility and ability to replicate natural tissue environments make them particularly valuable, especially in conjunction with TPP [24, 25]. Hydrogels can be classified into three main groups: natural, modified natural, and synthetic. Hydrogels have low photon absorption and a lack of photoinitiators; thus, natural hydrogels are not suitable for TPP. This led to the development of modified natural hydrogels, which are created by adding acrylates or methacrylates to natural hydrogels. Finally, epoxies form a class of thermosetting polymers known for their excellent mechanical properties, chemical resistance, and adhesive qualities. SU-8 is the most well-known epoxy-based photoresist, thanks to its wide use in microfabrication. Among the various characteristics of SU-8 photoresist, the most appealing are its high resolution, mechanical strength, transparency in the visible spectrum, and high resistance to solvents [24–26]. Moving on to hybrid materials, these are some of the main materials used in the TPP printing process. The most popular are the hybrid organic–inorganic photoresists. Examples of materials used so far include metal-containing materials, inorganic dielectric materials, and light-emitting materials. The popularity of these materials stems from the interesting properties of both their organic and inorganic parts. While the former contribute to their mechanical properties and porosity, the latter contribute more to thermal and mechanical stability.

4.2 Diffractive optical elements

Modifying the initial optical properties of light beams is crucial in many optical systems and applications. This allows for the customisation of incoming light beams for specific purposes, including filtering, focusing, and phase or amplitude beam shaping. So far, various methods have been employed to achieve these functions, including bulky optics, spatial light modulators (SLMs), and advanced micro- and nanoscale structures developed through electron beam lithography (EBL), focused ion beam (FIB), and 3D direct laser writing (DLW). However, key factors for future optical applications rely on miniaturisation, cost reduction, simplicity, and system integration. Thus, TPP has the potential to tackle these challenges through maskless printing, single-step processes, precise fabrication, and cost-effective methods.

Diffractive optical elements (DOEs) are components that utilise microstructured surfaces to manipulate light through diffraction. This differs from traditional optics, which depends on the surface shapes of components to manipulate or bend light. DOEs work by precisely manipulating the phase of incoming light through microstructures etched or fabricated onto their surfaces. These structures introduce varying optical path lengths, leading to diffraction and thus altering the wavefront. The amount of phase shift ($\Delta\phi$) at any given point (x,y) on a DOE's surface is given by:

$$\Delta\phi(x, y) = \frac{2\pi}{\lambda}(n - n_{\text{medium}})t(x, y), \tag{4.1}$$

where n and n_{medium} define the refractive index of the DOE material and that of the medium, respectively, and $t(x, y)$ defines the thickness required at each point to achieve the target phase modulation at a specific wavelength, λ.

DOEs can be grouped into different types, according to what they do to the incoming light beam. Examples include beam shapers, lenses (including diffractive lenses and Fresnel lenses), beam splitters, vortex and Bessel beam generators, gratings, and others.

Advancements in AM, especially in high-resolution TPP, have led to the development of various studies related to the fabrication of DOEs. One example is the 3D printing of refractive phase plates used to correct aberrations in multilayer Laue lenses (MLLs) [27]. These lenses have the potential to achieve numerical apertures in the hard x-ray regime, enabling diffraction-limited hard x-ray imaging with focal spot sizes as small as 1 nm. However, due to inaccuracies in the lenses' manufacturing process and technical limitations for high-NA x-ray lenses, they tend to exhibit optical aberrations and consequent poor performance. The addition of extra optical elements, such as a corrective phase plate with a precise inverse phase map, can suppress these malfunctions. Although some phase plates have already been tested, aberrations still appeared, and the diffraction limit was not ideal [28, 29]. Therefore, a highly adaptable, tailor-made refractive phase plate has been proposed for a crossed-pair MLL through the implementation of TPP direct laser writing (TPP-DLW). After fabrication, the phase plate (see the scanning electron microscope (SEM) image in figure 4.1(b)) was placed downstream of the horizontal and vertical MLLs, allowing for aberration correction [27].

To verify the performance of the printed refractive phase plate, the focusing performance and the residual wavefront error were measured with and without the refractive phase plate. The results revealed a residual wavefront error, with a reduction from 0.27λ to 0.17λ observed upon the addition of the refractive phase plate. Regarding the focal spot size, an enhancement in the horizontal full width at half maximum (FWHM) was achieved, showing a decrease from 33 to 25 nm, while the vertical FWHM remained the same throughout the entire process. Finally, it was also verified that when the refractive phase plate was used, the MLL presented a relative intensity identical to that of an ideal lens.

Other DOE examples explored through AM include high-performance kinoforms. Kinoforms are high-efficiency phase lenses capable of focusing light like a

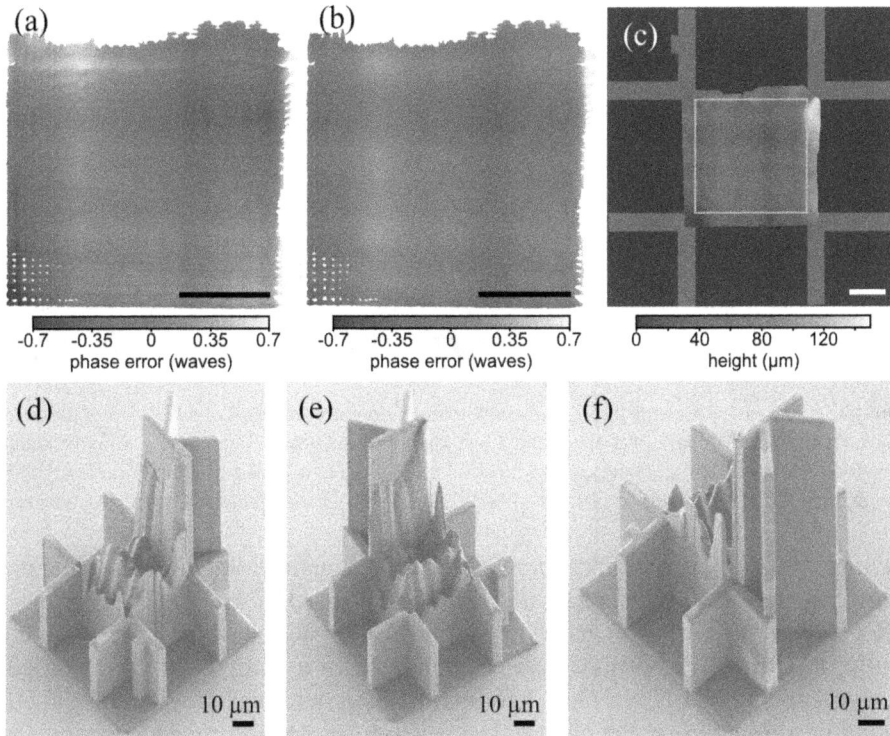

Figure 4.1. Wavefront error and phase plate design: (a) residual wavefront error; (b) modelled phase error to be compensated by the phase plate; (c) height profile of the phase plate obtained by expanding the wave field via convolution; (d–f) scanning electron microscope (SEM) images of the printed phase plate in different projections. Stabilising elements are included on the tallest phase plate structures. Reprinted with permission from [27]. © 2022 Optica Publishing Group.

traditional lens, but with nearly all the light going into the first diffractive order. Interest in these lenses has grown due to the increasing popularity of x-ray microscopy and its applications [30]. This enthusiasm for x-ray microscopy has sparked further interest in x-ray optics, as already illustrated by the production of x-ray Fresnel zone plates using subtractive methods [31, 32]. Despite the high resolution of these components, they have low diffraction efficiency. Therefore, other solutions have been proposed, namely, coherent diffractive imaging techniques. Kinoform lenses appear to be a promising solution, primarily due to their ability to efficiently perform phase modulation and their high focusing efficiency. So far, the methods commonly used to manufacture kinoform lenses have limitations regarding lens profile and efficiency [33, 34]. TPP technology appears to be an opportunity for the development of these lenses. An example of this can be found in [35], which reported the fabrication of a kinoform lens for advanced x-ray optics using the TPP method, as shown in figure 4.2. Results revealed focusing efficiencies greater than 15% for lenses of different thicknesses, measured at energies between

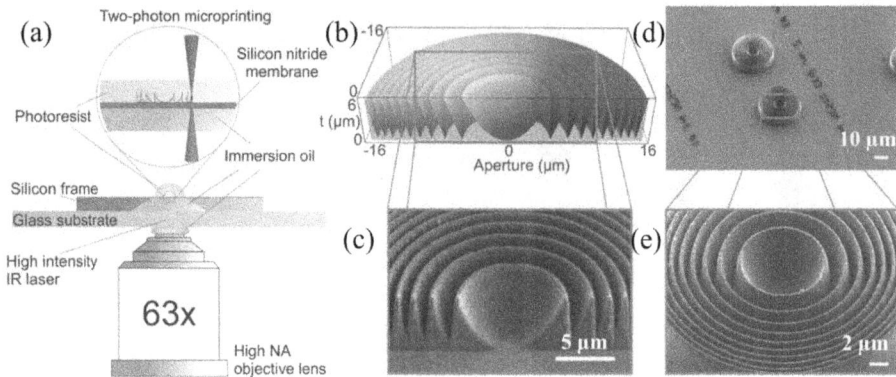

Figure 4.2. (a) Schematic showing the TPP fabrication of the kinoform lens. (b) Cut-away view of the designed kinoform lens. (c) SEM image of a nanoprinted half-kinoform lens. (d) A portion of an array of kinoform lenses printed with different parameters. (e) Magnified view of the central part of a kinoform lens [35] John Wiley & Sons. © 2018 The Authors. Published by WILEY-VCH Verlag GmbH & Co. KGaA, Weinheim.

900 and 1800 eV, which accounted for approximately 95% of the calculated theoretical efficiency, showcasing the potential of these lenses.

Recently, research has focused on AM for longer wavelengths, namely on a flat nanofocaliser that transforms a laser beam into a subwavelength spot array based on the fractional Talbot effect [36]. The interest in generating these flexible, focused spotlight arrays relies on the potential to perform optical sorting and manipulation. For this to happen, delicate modulation of the phase delay of the incident wavefront is required, and this involves the use of curved lenses arranged periodically, which is rather complex to achieve through traditional techniques. Thus, the study reported in [36] took advantage of the TPP-DLW technique, reporting the fabrication of a flat nanofocaliser. Experiments were conducted at a wavelength of 750 nm using a nanofocaliser consisting of periodic unit elements with dimensions of 300 nm (width) × 600 nm (length) × 585 nm (height), resulting in the formation of a 5 × 6 light spot array, with each spot having an FWHM of ≈0.82 λ. This paved the way for the realisation of subwavelength optical devices that can be readily integrated into existing optical systems.

The environmental issues related to humidity and temperature, as well as the long-term stability and intended use of 3D phase elements, can be problematic if TPP structures are fabricated using organic photopolymers. To address this, Lightman *et al* [37] employed a sol-gel hybrid material to produce 3D phase micro-optic elements with more glassy properties than purely organic ones. The obtained 3D-printed elements were designed to shape incoming Gaussian beams into beams with flattened intensity distributions and either square or line profiles. Using this approach, the authors were able to print micrometre-scale refractive phase elements with superior mechanical properties and handle higher power laser damage thresholds compared to those of purely organic optical devices.

Further important research related to the use of DOEs was the generation of optical vortex beams. Optical vortex beams are laser beams with a helical phase front, meaning that as the beam propagates, the wavefront rotates around its central axis. Their azimuthal phase dependence is $\exp(il\phi)$, where l is an integer representing the topological charge, and ϕ is the angular coordinate. These beams can carry an orbital angular momentum (OAM) of $l\hbar$ per photon, where $\hbar$ is the reduced Planck constant. These beams attract particular interest for several applications, including optical tweezers, quantum optics, and optical communications [38–40]. This plethora of applications has already led to reports of optical vortex beams in different spectral regions, such as the visible [41], infrared [42], and long millimetre wavelengths [43]. In the field of communications, they can be used in mode-division multiplexing applications, since beams with different l values are orthogonal; thus, they can be spatially multiplexed and demultiplexed on the same physical channel. Therefore, each beam can be used as a carrier of information, allowing an increase in data transmission capacity. Currently, most of the methods used to sort modes (demultiplexing) employ bulky and large devices. Several approaches have already been explored for use in demultiplexers, including systems based on two SLMs [44], large-scale elements manufactured by CNC milling [43] or diamond turning [45], and diffractive elements produced through EBL [46]. While these techniques performed well in controlled environments such as laboratories, they required other components, such as lenses or mirrors, making it challenging to integrate these mode sorters into optical systems such as the tips of OFs. This limitation prompted a search for other solutions. The solution was to directly manufacture (through TPP-DLW) a mode sorter of vortex beams, as demonstrated in [47]. For their first approach, the authors developed a demultiplexer consisting of two diffractive phase elements and presented their results. In their second approach, they utilised an integrated and fully aligned system that combined the two diffractive elements into one compact and stable device. The printed mode sorter was able to distinguish between pure and mixed OAM states in the range of $-3 \leqslant l \leqslant 3$ within a bandwidth range of 300 nm (690–990 nm), which was limited by the dispersion of the photoresist [47].

For longer wavelengths, such as the mid-infrared, the most commonly adopted methods to convert an incoming Gaussian beam into an optical vortex beam include spatial light modulation [48] and silica-based spiral-phased plates [49]. However, these are not transparent in the mid-infrared region, an important part of the electro-magnetic spectrum, as it corresponds to the fundamental vibrational and rotational energy transitions of several molecules. Thus, taking into account the potential of the TPP-DLW technique and the possibility of using mid-IR-transparent polymer resin, Zhou et al reported the additive manufacture of a 120 μm polymer-based spiral phase plate consisting of 32 well-defined segments [50], as shown in figures 4.3(a) and (b). The height of these elements was defined to increase linearly until it reached a maximum height difference of 6.2 μm, allowing the creation of a 2π helical phase change. The polymer resin used for the 3D printing offered low attenuation in the region between 2500 and 5500 nm, and the fabricated structures were able to generate a perfect vortex beam (see figure 4.3(b)), as idealised numerically.

The additive manufacture of beam generators is not limited to free-space optics, as demonstrated in the examples above. Beam generators can also be fabricated at

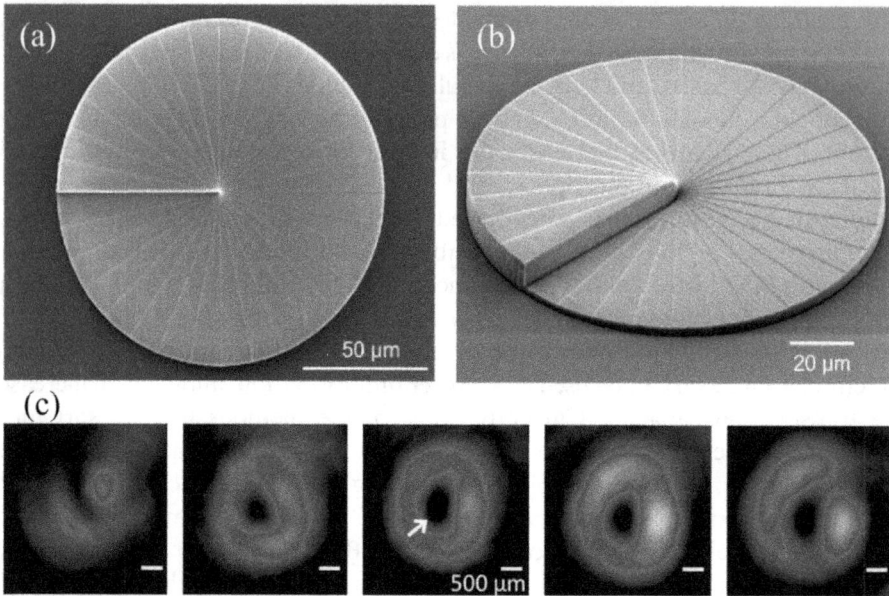

Figure 4.3. SEM images: (a) top view and (b) side view. (c) Images of the mid-IR vortex beam created by the spiral phase plate at different wavelengths; from left to right: 2.9, 3.1, 3.3, 3.5, and 3.7 μm. A sharp doughnut shape with a uniform ring is visible at 3.3 μm. Reprinted from [50], Copyright (2022), with permission from Elsevier.

the tip of an optical fibre (OF). This has already been demonstrated through the direct 3D printing of a kinoform spiral zone plate (KSZP) at the tip of an OF [51], as seen in the SEM images of figures 4.4(a)–(d).

The plates presented a spiral continuous surface relief structure that converted the propagating beam in the OF into a single-focus vortex beam (see figures 4.4(e)–(h)). The structure thus created exhibited a focusing efficiency of more than 60% and a vortex beam purity of 87% [51]. These results showcase the excellence of the focus conversion efficiency of the 3D-printed KSZP, highlighting its potential for low-power fibre devices in optical communication and optical manipulation applications.

The generation of Bessel beams through the use of phase plates produced through AM is just another example of the potential of TPP printing [52]. When illuminated by light with a wavelength of 1550 nm, the phase plates indicated that the profile of the Bessel beams remained invariant for distances ranging from 0 to 800 mm. These results demonstrate that the 3D-printed phase plate can generate non-diffracting Bessel beams, which have potential applications in integrated optics.

Still on the topic of the additive manufacture of DOEs, a study by Lightman *et al* [53] examined direct printing on the facets of OFs. In this study, the authors manufactured a helical axicon micro-optical element, as well as a parabolic lens at the tip of a single-mode fibre (SMF). The helical axicon was designed to generate a high-order Bessel beam carrying an OAM value of $1\hbar$ per photon, while the

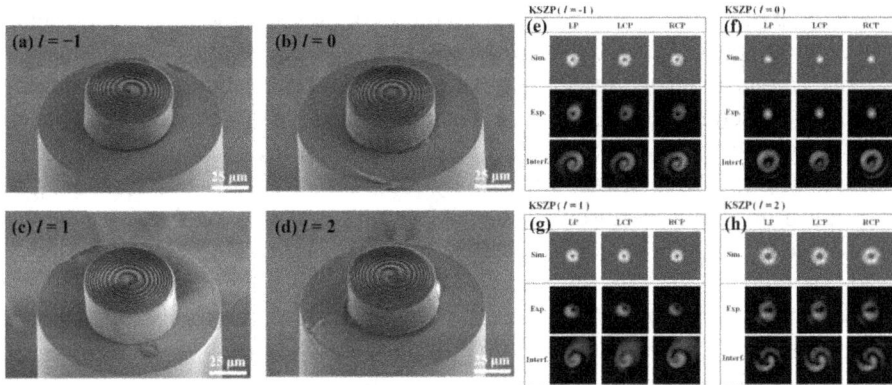

Figure 4.4. (a–d) SEM images of the TPP KSZP fabricated at the tip of an optical fibre (OF), for different values of topological charge. Panels (e)–(h) show the corresponding focal spot profiles. The top row, middle row, and bottom row images correspond to the numerical results, experimental measurements obtained through direct observation, and experimental measurements obtained through interference patterns, respectively. Adapted with permission from [51]. © 2020 Optical Society of America.

parabolic lens was added to compensate for the spherical phase front of the diffracted beam at the OF exit side.

The devices shown above clearly illustrate that AM is a viable alternative to traditional DOE fabrication methods, sometimes resolving existing challenges. This also highlights AM's potential for application in other areas, such as diffractive neural networks [54], where significant advancements have already been made. Utilising AM's capacity to create intricate, free-form, multi-material structures with submicron accuracy, future optical systems can be customised for unparalleled functionality, compactness, and integration in both classical and quantum domains.

4.3 Imaging optics

AM is revolutionising the field of imaging optics by enabling the production of complex and highly customised optical components with unprecedented precision and efficiency. This innovative technology enables the manufacture of intricate designs that were previously impossible, time-consuming, or cost-prohibitive using traditional manufacturing techniques. In imaging optics, AM facilitates the creation of lightweight, compact, and highly integrated optical systems, including lenses and mirrors. Not surprisingly, the development of AM technologies has led to the capability to 3D print components at smaller scales than ever before. Imaging optics was no exception, having already explored the manufacture of optical components at the micro- and nanoscales.

Currently, lenses can be classified into three main categories, depending on their properties: refractive lenses, diffractive lenses, and metalenses. Refractive lenses can be subdivided into conventional millimetre-scale bulk lenses and microlenses, depending on their physical size and application domain. Since conventional bulk lenses were explored in chapter 2, this chapter focuses on microlenses, examining recent developments and fabrication strategies.

Today, various microscale lenses have been produced through AM technology, including both concave and convex designs. Their first descriptions appeared decades ago with the use of different technologies, such as the photoresist reflow technique [55], grey-tone photolithography [56], DLW with a UV laser [57], and deep x-ray lithography exposure to a moving x-ray resist [58]. All these methods were capable of manufacturing inexpensive microlenses but struggled to achieve the desired geometries at the microscale. Consequently, different solutions were explored, one of which was the use of fs TPP. This technique was selected because it can produce structures with arbitrary designs at a lateral spatial resolution of about 100 nm [7] and can manufacture components at specific locations, such as the tip of an OF [59]. However, the design of microlenses requires a smooth, high-quality surface. Considering the spherical surface of lenses, this can become challenging in TPP fabrication if one uses the parallel linear scanning mode, i.e. the laser scans in straight, parallel lines, typically across 2D planes. To solve this, in [60], an annular scanning mode was used in conjunction with a continuous variable scanning step space (see the schematic in figure 4.5), enhancing the surface quality of a 2×2 microspherical lens array 15 μm in diameter and a micro Fresnel lens with a 17 μm diameter.

The design freedom offered by TPP-DLW also led to the development of aspheric microlenses [19]. These lenses have the potential to eliminate the spherical aberration commonly observed in spherical lenses, which is often characterised by blurry images, a result related to the different focal points formed by different rays travelling parallel to each lens's principal axis. This can typically be solved using complex multilens systems. However, the advent of TPP-DLW enables the printing of complex shape profiles; thus, the additive manufacture of a single aspheric lens presents an opportunity to reduce complexity, size, and weight. The results presented in [19] were achieved for the additive manufacture of a parabolic lens profile with a radius of 10 μm and a height of 10 μm, fabricated on a SU-8 photoresist film. When compared with theoretical results, the authors reported relative errors of 0.2% and

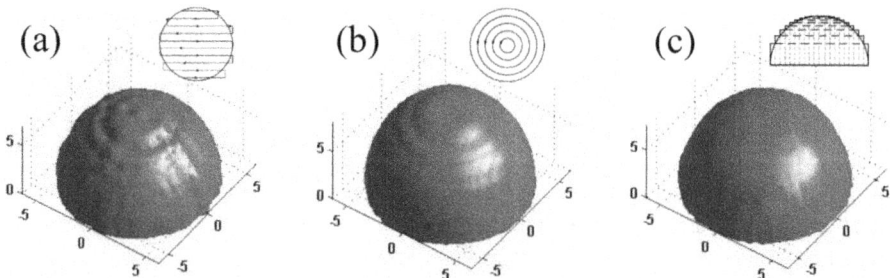

Figure 4.5. Numerical results for microlens surface quality based on: (a) a parallel linear scanning method with a fixed constant delta Z, (b) an annular scanning method with a fixed constant delta Z, and (c) an annular scanning method with a dynamical Z_i, where $Z_i = [R^2 - (i \cdot lx)^2]^{1/2}$. Reprinted with permission from [60]. © 2006 Optical Society of America.

an overall fabrication precision of 20 nm. These results were attributed to the small size of the voxels (<100 nm), the low polymerisation shrinkage of the photoresist, the high-precision moving stages (1 nm resolution), and the small voxel overlap. To further extend the results of this study and to fully take advantage of the laser technology, the authors fabricated a lens array with a 100% filling ratio. In this way, it was possible to utilise the entire incident light and enhance the image brightness of liquid crystal displays, as well as raise the out-coupling efficiency in organic light-emitting diodes. This was done for lenses with a radius of 5 μm and a height of 1.5 μm. To enhance the surface roughness of the printed lens array, the authors also used the annular scanning mode reported in [60], allowing them to achieve better results than those available using the common raster scan.

While the development of aspheric microlenses with low curvature profiles allows for the reduction of aberrations, as confirmed by the AM microlenses described in [19], the second surface of these lenses is flat; thus, adjustment of the refraction is only possible through the first surface. Yet, the significant difference in the optical path between the central region and the lens edges introduces aberrations. To compensate for this and to reduce the aberration, [61] proposed the TPP-DLW fabrication of a concave–convex microlens. The lens had a 30 μm diameter and radii of curvature of 30.1 and 15.4 μm and was fabricated on an SU-8 film photoresist, giving a theoretical focal length of 49.1 μm. The lens exhibited optimal optical performance, with a reported focal length shift of approximately 4.6% across the wavelength span from 450 to 660 nm, indicating potential applications in imaging and near-field recording systems.

Simpler does not always mean efficient; thus, complex optical imaging systems are often required, making the stacking of multiple singlet lenses a necessity. Several methods have already been developed to manufacture small, complex, and high-performance micro-optical systems [62–69]. However, limitations regarding minia-turisation, design freedom, and precision alignment capabilities have led to the development of studies dedicated to the additive manufacture of multi-microlens systems. Gissibl *et al* [70] were pioneers in demonstrating the TPP-DLW of different multilens objectives. Figure 4.6 shows examples of the three nonspherical-surface lenses (singlet, doublet, and triplet) reported in their study.

The lenses presented in [70] were designed to check the effect of the number of refractive interfaces on the field-dependent optical aberrations, namely, field curvature, coma, astigmatism, and distortion. These lenses consisted of one surface for the singlet to up to five for the triplet lens seen in the SEM images shown in figure 4.6(c). The lens design specified a large field of view (FOV) of 70°, a 120 μm diameter, and compound heights between 100 and 200 μm. The compound lenses were printed in a single run using IP-S photoresist on top of a coverslip. An outer shell included in the lens design served as a support for the lens and guaranteed lens alignment both axially and longitudinally. The simulations shown in figure 4.6(a) revealed that the focusing performance across different wavelengths improved as the number of surfaces increased. The results were corroborated by simulating the 1951 USAF resolution test chart seen in figure 4.6(b) and by the experimental results shown in figure 4.6(d), where a performance enhancement can be observed for the

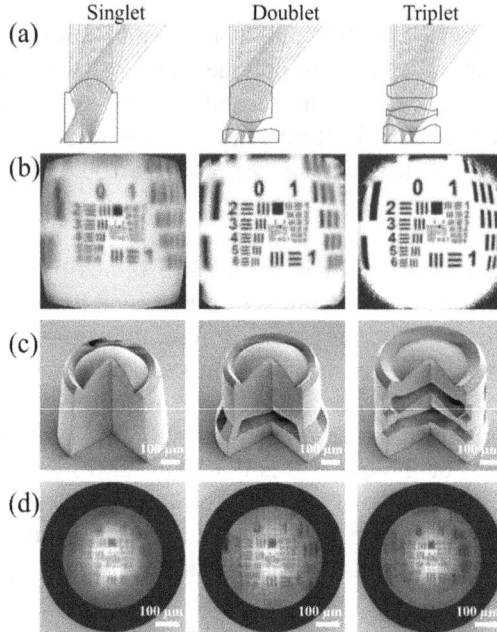

Figure 4.6. (a) Optical lens designs carried out using ZEMAX. (b) Simulated images of the 1951 USAF resolution test chart. (c) SEM images (cut-away view) of the different manufactured lens structures with a total height of $\approx 115\,\mu m$ and a focal length of $63.3\,\mu m$. (d) Images of the 1951 USAF test chart observed through the different printed lenses. Scale bars are $20\,\mu m$. Reproduced from [70], with permission from Springer Nature.

triplet lens, while barrel-like distortion at the edges of the image and nonuniform magnification are clearly visible in the case where a single lens is used. Tests of the chromatic aberration behaviour showed similar performance across the three sets of lenses under study, and to compensate for this, the authors proposed the use of the classical achromatic lens design consisting of two materials with different Abbe numbers. As proof of concept, the authors also demonstrated the capability to directly manufacture doublet lenses on a complementary metal–oxide–semiconductor (CMOS) imaging sensor and a triplet lens on the tip of an imaging OF intended for endoscopic applications, as depicted in figure 4.7.

The application examples of the microlens objectives presented in [70] yielded high-quality detail imaging results as shown in figures 4.7(c) and (f) and good repeatability (figures 4.7(b) and (c)). The potential of these results opened opportunities for a new generation of ultracompact optical elements for endoscopic instruments and miniaturised microscopes to be used in medical engineering, as well as optical lenses directly installed on CMOS sensors for miniaturised robots and drones. An example of such work was reported one year later by the same team for the manufacture of a multi-aperture foveated imaging system. Foveated vision, also called eagle-eye vision, refers to a visual system, biological or artificial, where spatial resolution is highest at the centre of gaze (the fovea) and decreases toward the periphery. This is usually found in animals, especially among predators. However,

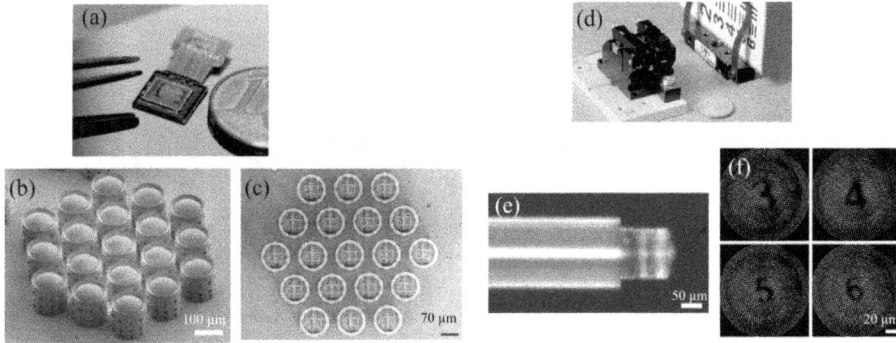

Figure 4.7. TPP-DLW of microlenses (doublet arrays (b) and triplet lens (e)) on top of a CMOS image sensor (a), and an imaging OF (e). Image of a 1951 USAF resolution test chart (group −2) captured by the microlenses on the CMOS sensor at a distance of 3 cm (c). Measurement setup of the imaging fibre containing a triplet lens, positioned 3 mm from the 1951 USAF test target (d), and its corresponding image of the notations of the elements of group 0 (element 6 has a height of ≈1.6 mm). Reproduced from [70], with permission from Springer Nature.

foveated imaging is also very important in applications such as machine vision, medical imaging, virtual reality, and robotics, where full resolution everywhere is expensive. Thus, in order to demonstrate a foveated system, the authors 3D printed a multi-aperture design that combined four aberration-corrected air-spaced doublet objectives with different focal lengths (equivalent focal lengths for a 35 mm film of $f = 31$, 38, 60, and 123 mm) and a common focal plane situated on a CMOS image sensor. The lenses were printed in a miniaturised configuration arranged in a 2 × 2 pattern and occupying an area equivalent to 300 μm × 300 μm, while their height was ≈200 μm, making this configuration one of the smallest reported in the research literature. Moreover, the system presented good image quality and functionality.

Tunable-focus liquid crystal (LC) microlens arrays are crucial optical elements for image processing, beam steering, wavefront correction, and switchable 2D/3D application displays. To adjust the focal length, it is essential to develop a parabolic phase profile through spatial manipulation of the LC director's reorientations. This can be achieved by applying an inhomogeneous electric field distribution or establishing a gradient pretilt angle alignment. To date, different methods have already been suggested to achieve this, such as patterned electrodes [71], polymer stabilisation [72], composite alignment [73], and composite lenses [74]. Regarding composite lenses, it is essential to employ a reliable method to align the LC on the polymer microlens array. To achieve this, the study described in [75] used TPP-DLW to fabricate a parabolic 16 × 16 microlens array (each with an area of 120×120 μm^2) on an indium tin oxide (ITO)-coated planar substrate. This enabled the generation of the desired inhomogeneous electric field for the lens and also served as a homogeneous-type LC alignment layer (see figures 4.8(a)–(d)). The results showed that the phase profile and, consequently, the focal length could be dynamically tuned by applying a voltage. The focusing properties revealed well-defined focal points, and the imaging capabilities of the composite adaptive

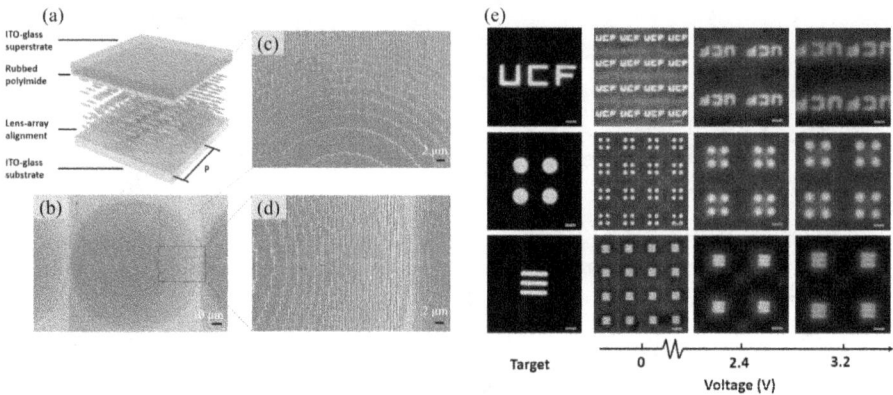

Figure 4.8. (a) Schematic of the 16 × 16 microlens array. (b–d) Corresponding SEM images of the TPP-DLW microlenses. The red arrows highlight the nano-groove directions. (e) White-light imaging of three targets (rows) as a function of applied voltage (columns). From top to bottom: the letters 'UCF', a four-dot target, and a three-bar target. The red scale bar is 20 μm. Adapted with permission from [75]. © 2018 Optical Society of America.

microlens array proved to function well, as can be observed in the microscope images of the imaging targets shown in figure 4.8(e) [75].

The microscopic images of the imaging targets shown in figure 4.8(e), experimentally taken in [75], were captured with white light using the microlens array under different applied voltages. Without an applied voltage, the images are virtual and erect, since the microlens array is concave. Conversely, they become real and inverted with an applied voltage because of the convex microlens array shape. The capability to resolve a three-bar target with 57.0 lp mm^{-1} resolution was demonstrated, and the array was also able to separate a four-dot target.

The compound eyes (CEs) found in insects (see figure 4.9(a)) are advanced and complex imaging systems composed of several packed and hemispherically distributed ommatidia, working independently as photosensitive units and cooperating with each other to realise prey recognition and enemy defence. These structures have attracted special interest due to their remarkable characteristics, such as small size, distortion-free imaging, wide FOV, and high-sensitivity movement tracking ability, making them an inspiration for the development of artificial CEs. As a result, solutions such as the implementation of a microlens array in commercial CMOS detectors [76] have already been proposed, allowing for high imaging resolution but a limited FOV due to the planar structure of the imaging sensor (see figure 4.9(b)). The design flexibility provided by TPP-DLW offers unprecedented opportunities, and this has already led to the development of the TPP fabrication of an optoelectronic CE (see figure 4.9(e)) [77].

The research presented in [77] included the manufacture of polymer CEs comprising 19–160 ommatidia with logarithmic and spherical profiles. As expected, spherical ommatidia showed low performance, presenting defocusing problems. On the contrary, logarithmic ommatidia enabled a high depth of field and focus range, allowing for their direct integration into a commercial CMOS detector (see the

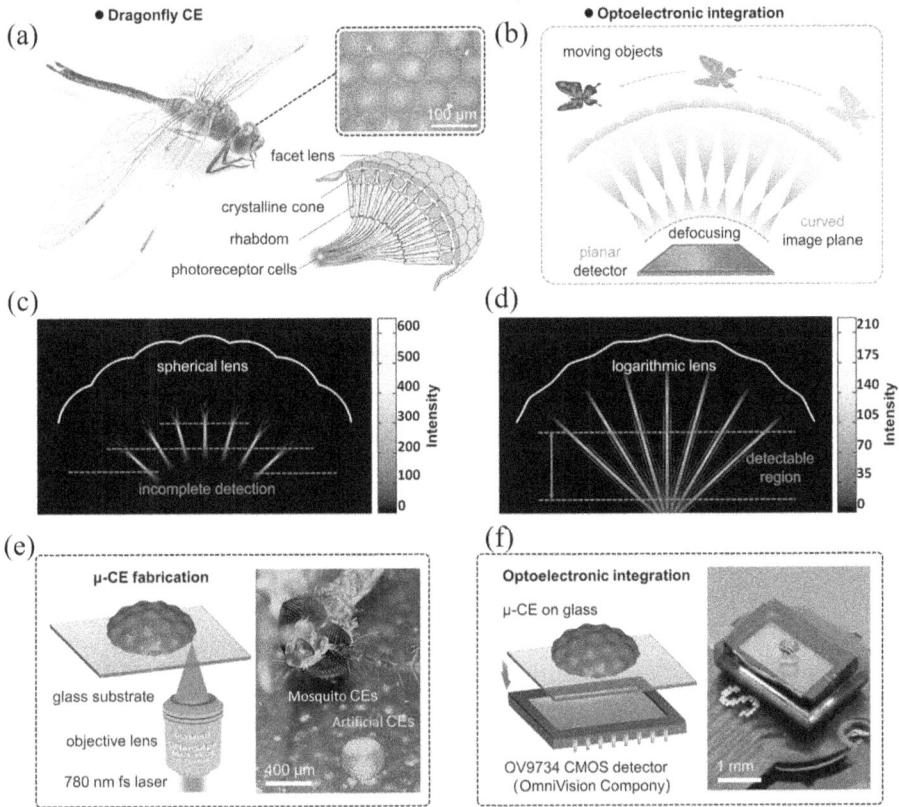

Figure 4.9. Design and fabrication of an optoelectronic CE camera. (a) Photograph of a biological insect eye with its corresponding CE micrograph and a schematic of the CE structure. (b) Problem associated with the integration of a common CE spherical lens (i.e. the presentation of a curved image plane) and a CMOS detector with a planar surface. Simulated light field distribution of CEs with: (c) spherical ommatidia and (d) logarithmic ommatidia. Schematic of the (e) TPP-DLW fabrication and (f) CMOS optoelectronic integration of the CE. The insets of (e) and (f) show a comparison between a natural CE and an artificial CE and the integration of an artificial CE into a CMOS detector, respectively. Reproduced from [77]. CC BY 4.0.

numerical result comparisons in figures 4.9(c) and (d)). The optoelectronic integrated micro-CE camera (see figure 4.9(f)) enabled imaging with a large FOV (90°), spatial position identification, and sensitive trajectory monitoring of moving targets [77].

Another interesting study of CEs that employed AM technology was conducted by Dai *et al* [78], who proposed the manufacture of a biomimetic apposition CE. The manufacturing process consisted of two steps: the fabrication of a mould using a projection micro-stereolithography printer and a second step that involved the manufacture of the microlens using a microfluidic-assisted moulding technique. A representative schematic of the compound eye, its manufacturing process, and SEM images can be observed in figure 4.10.

Figure 4.10(d) shows an SEM image of the biomimetic CE [78]. This has a radius of 2.5 mm and contains 522 microlenses, each with a diameter of 180 μm,

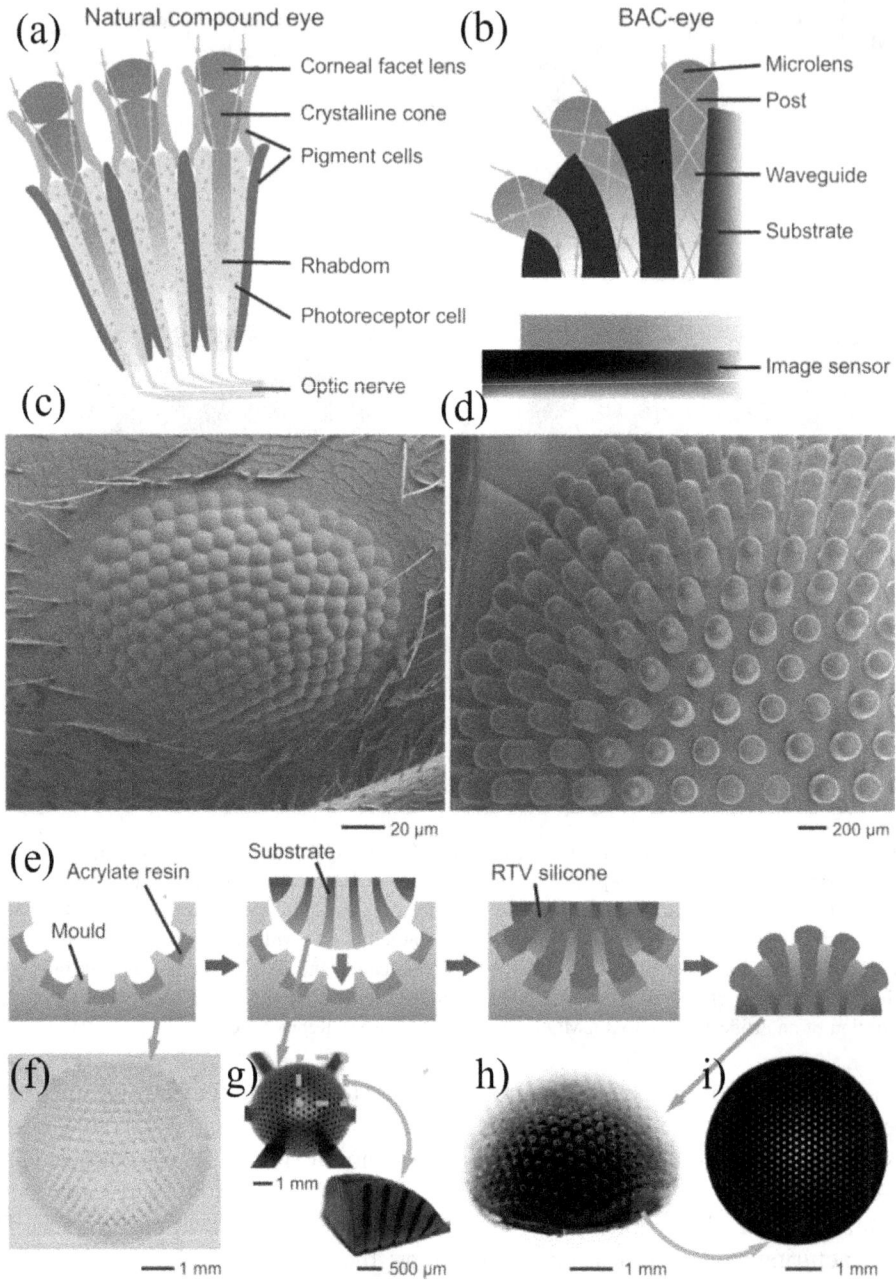

Figure 4.10. (a) Schematic structure of a biomimetic CE and (b) its cross-section. (c) SEM image of the eye of an Asian needle ant. (d) SEM image of the biomimetic CE. (e) Biomimetic CE fabrication steps. (f) 3D-printed mould. (g) 3D-printed substrate and one of its sliced sections. (h) Image of the biomimetic compound eye after removal from the mould. (i) Image showing the flat bottom region of the biomimetic CE. Reproduced from [78]. CC BY 4.0.

hexagonally and omnidirectionally distributed across its hemispherical dome. The most peripheral ommatidia are oriented at ±85° with respect to the vertical axis, extending the viewing angle of the CE to 170°. Each microlens is connected to the eye's bottom surface through refractive-index-matched waveguides, mimicking the natural rhabdoms. Full-colour panoramic views and position tracking were achieved by placing the eye on a commercial imaging sensor, highlighting potential applications in endoscopy and robotic vision.

It is not possible to observe small features (smaller than $\lambda/2$) with a classical optical microscope due to the Abbe diffraction limit. This originates from the exponential decay of evanescent waves with high spatial frequencies. To overcome this issue, several technologies have already been reported, such as superlenses [79], scanning probe microscopy [80], and fluorescence microscopy [81]. However, the use of dielectric microstructures coupled with a conventional optical microscope provides a more straightforward and cost-efficient way to achieve super-resolution imaging [82]. It has been verified that a microstructure geometry is vital for super-resolution imaging [83]. Therefore, the use of TPP-DLW appeared to be a suitable candidate for this purpose. To verify this, in the study described in [84], four types of microstructures, namely a cylinder, a truncated cone, a hemisphere, and a protruding hemisphere, were designed on a semiconductor chip substrate. These achieved super-resolution imaging in the optical far-field region, with the possibility of resolving features as small as 100 nm under a conventional optical microscope.

The TPP-DLW manufacture of micro- and nanoscale compound lenses with intricate parts, fabricated in a reproducible way, has achieved unprecedented scales and results. However, lenses such as the ones shown in [70] rely on a single material's refractive index, which limits the degrees of freedom in light manipulation. The authors of [70] proposed the TPP printing of achromatic lenses made of different materials with different Abbe numbers to reduce spherical and chromatic aberration. However, the details of how to perform the material swapping during the printing process were not described. The ability to control light guidance through TPP-DLW printing using geometric free-form shapes offers unprecedented design freedom. Yet, the possibility of adding new degrees of freedom in terms of refractive index manipulation along the 3D-printed parts is very appealing, for instance, in applications related to anamorphic light beam shaping, where aberration-free and plane optical surface components are desirable [85]. The first study exploring this possibility was reported in [85], where the refractive index change was achieved through the control of laser dosage, specifically by adjusting the laser velocity. This allowed the authors to report a gradient refractive index (GRIN) structure, with a maximum refractive index change on the order of 10^{-2}. Later, another group reported a subsurface controllable refractive index via beam exposure (SCRIBE) [86], enabling the refractive index to change over a more extensive range, i.e. 0.3. In this study, the authors spatially tuned the refractive index of the printed components with submicron resolution by adjusting the laser power during laser writing, thereby altering the photoresist fill fraction and, consequently, the refractive index and dispersion. The several interesting optical components printed through SCRIBE included singlet and doublet lenses designed to control chromatic dispersion. Examples of these lenses are shown in the multiphoton fluorescence microscopy images presented in figure 4.11(a). The most notable result was obtained for the achromatic doublet shown on the right side of figure 4.11(a), which was 3D printed simultaneously by varying the laser exposure. This eliminated the requirement for multiple photoresins and writing sequences. Using this technique, the authors designed the curvature radius and Abbe numbers of the two lenses to manipulate the chromatic aberration curve while still maintaining a small chromatic focal shift (≈ 1.9 µm).

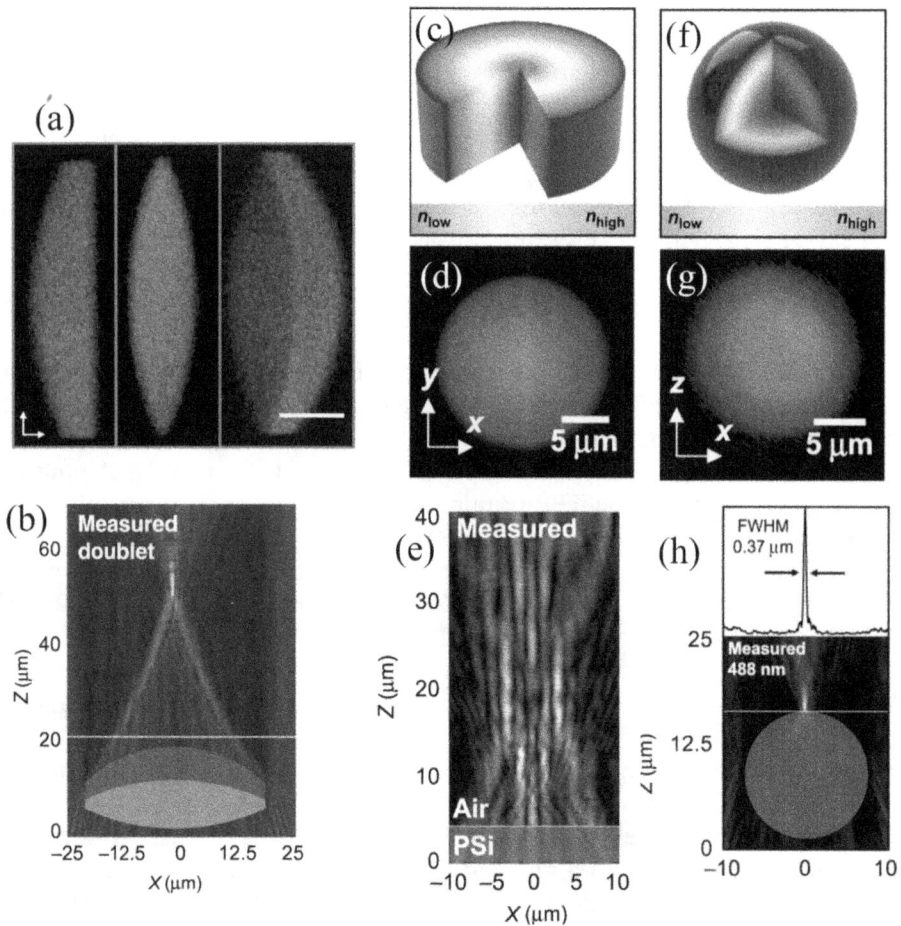

Figure 4.11. Gradient refractive index (GRIN) lenses. Multiphoton fluorescence microscopy images of the chromatic dispersion control lenses, from left to right: plano-convex, biconvex, and doublet lenses. In (b), the focal behaviour of the doublet lens is shown. Panels (c) and (f) are representations of the refractive index distribution of the GRIN lenses, namely a planar axicon lens and a spherical Luneburg lens, respectively. Panels (d) and (g) are the corresponding multiphoton fluorescent images of each lens's cross-section, while panels (e) and (f) correspond to their focal behaviour. Reproduced from [86]. CC BY 4.0.

The SCRIBE technique [86] was also explored for the fabrication of GRIN lenses, which can reduce geometric aberrations. This makes them an alternative to conventional geometric optical lenses. The fabricated GRIN lens had a 20 μm diameter with a radial refractive index decrease as depicted in figure 4.11(c), reducing from 1.8 to 1.6 at 633 nm. This refractive index variation can be visualised through the fluorescence contrast intensity visible in the multiphoton fluorescence microscopy image in figure 4.11(d). The performance of this planar axicon was tested, showing a Bessel-like beam with a ring-shaped intensity distribution when illuminated with a plane wave. This study was later extended to the fabrication of the smallest (i.e. 15 μm) spherical Luneburg lens (see the representation in figure 4.11 (f)). These lenses are aberration- and coma-free, presenting a spherically symmetric refractive index profile as seen in the multiphoton image of the GRIN lens midsection region (see figure 4.11(g)), where a gradual change in fluorescence intensity from the centre to the edge is observable. The focusing capability of the Luneburg lens was confirmed using visible wavelengths, and resolution-limited measurements at FWHMs of 0.37 and 0.41 μm were attained at 488 and 633 nm, respectively.

The majority of the AM micro-optics reported in the literature were fabricated using organic polymers based on acrylates, methacrylates, epoxies, or similar photopolymerisable monomers [87, 88]. However, depending on the application, constraints related to thermal stability, mechanical robustness, spectral range transparency, etc. can limit their deployment. On the other hand, silica glass exhibits several excellent material properties, including thermal, mechanical, and chemical stability, as well as a wide range of transparency across various wavelengths. While different AM techniques have been used to print silica glass, they tend to be challenging when addressing microscale features. However, the materials used rely on sol-gel methods, where different organic mixtures are loaded with 50 wt.% of silica nanoparticles. As a result, all these structures must be sintered at temperatures close to 1200 °C to obtain functional, fused, transparent silica glass free of organic compounds. Thus, the substrates on which micro- or nanostructures are created need to survive such environments, which restricts the range of applications. Additionally, the burnout of organic components during the sintering stage results in a high percentage of shrinkage and defects, including bubbles and hidden layers. To overcome these problems, the manufacture of silica glass micro-optics using hydrogen silsesquioxane (HSQ), an inorganic silica-like material widely used as a high-resolution negative-tone resist, has recently been reported. For this purpose, HSQ was selectively crosslinked into silica glass in 3D through a nonlinear absorption process by exposure to sub-picosecond laser pulses with a wavelength of 1040 nm [89]. Another strategy reported in the research literature was the use of a solvent-free, pre-condensed liquid silica resin (LSR) for the TPP fabrication of microglass optics. The synthesis was based on the acid-catalysed polymerisation of tetramethoxysilane together with a substoichiometric amount of water (an aqueous solution) and 6.5 mol% of methacryloxymethyltrimethoxysilane (MMTS) as the photocurable moiety [90]. Using this technique, a relatively simple structure based on a plano-convex microlens with a 25 μm radius of curvature was 3D printed on top of four pillars attached to a quartz glass substrate [90]. The results obtained for sintering processes at low temperatures (600 °C) showed shrinkage values as low as 17% and a surface roughness of less than 6 nm. Despite the successful results, the authors reported some limitations of their earlier achievement in a subsequent article [91]. Among these were the deformations incurred during the printing and thermal treatment processes, as well as the inability to print a three-lens objective with a high aspect ratio (a diameter of 50 μm and a height of 100 μm) due to the weak supporting structure. To improve their earlier work, the team optimised the LSR by increasing the crosslink points through the adjustment of the MMTS ratio during synthesis. From this, they observed that the best balance between the shrinkage of the printed component and its mechanical properties was achieved for 15% MMTS, allowing them to achieve a shrinkage value of 22%, i.e. slightly higher than in their earlier study and indicative of the higher organic material content. After the optimisation process, the new resin was tested in the manufacture of diverse micro-optical singlet spherical lenses, including plano-convex, plano-concave, and biconvex lenses, as well as free-form lenses such as aspherical plano-convex lenses and Fresnel lenses. Examples of these lenses can be observed in figure 4.12 [91].

The resolving power of the lenses was evaluated by visualising the images of a 1951 USAF resolution target produced by the lenses, as shown on the right-hand side of each SEM image in figure 4.12. The spherical aberration correction of aspherical versus spherical lenses is easily observed on the right-hand sides of figures 4.12(c) and (d), respectively, proving the capabilities of the design freedom offered by the produced lenses.

The authors of [91] also demonstrated the capability to produce complex imaging systems, such as a micro-objective consisting of three lenses; a microspectrometer

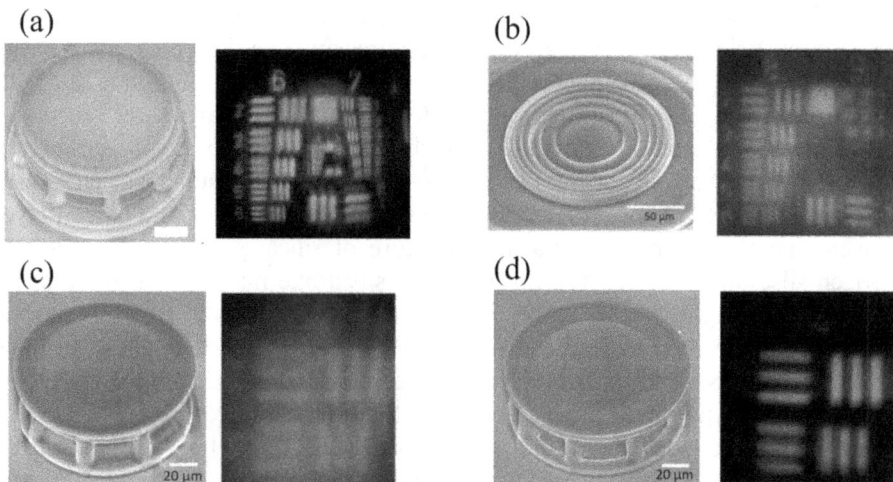

Figure 4.12. SEM images of glass micro-optical lenses and corresponding photos of a 1951 USAF resolution target captured by the lenses ((c) and (d) refer to the fourth group). (a) Spherical plano-concave lens, (b) Fresnel lens, (c) spherical plano-convex lens, and (d) aspherical plano-convex lens. Reproduced from [91]. CC BY 4.0.

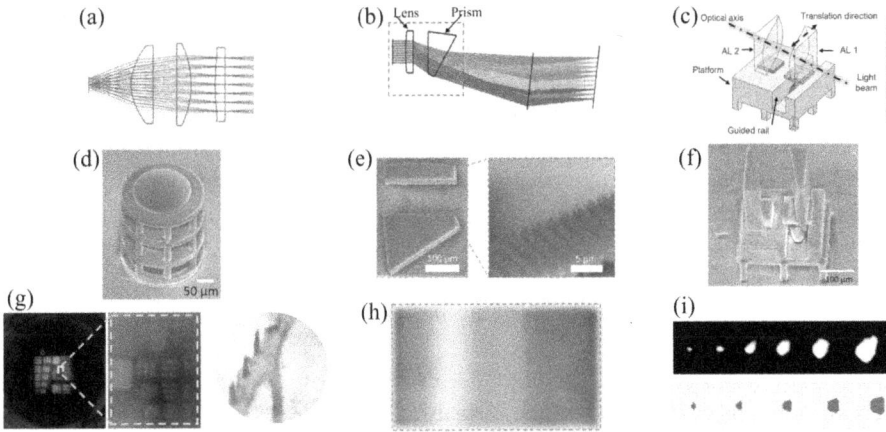

Figure 4.13. Precision TPP 3D printing of complex glass optical systems. Optical layouts of lens systems based on: a 3D-printed micro-objective with three elements (a), a microspectrometer with a printed dispersion assembly (b), and an Alvarez lens (c). The images in parts (d)–(f) correspond to SEM images, while those in (g)–(i) show the corresponding results for each optical system. Reproduced from [91]. CC BY 4.0.

composed of a lens with a grating on its surface, followed by a prism; and finally, a pair of Alvarez lenses. A schematic of each imaging system, together with SEM images of the lenses, can be seen in figure 4.13 [91].

The characterisation of the optical systems presented in figure 4.13 and reported in [91] revealed impressive results, where the micro-objective was able to: clearly resolve Element 3 in Group 9 of the 1951 USAF resolution test target (see figure 4.13 (g)), corresponding to a resolution of 780 nm; disperse a spectrum from a collimated white beam (see figure 4.13(h)); and change the focal spot size of a collimated beam (see the experimental and theoretical spot sizes in the top and bottom images of figure 4.13(i)) through the lateral displacement of one Alvarez lens over the other. To conclude, thanks to the properties of the glassy material, the 3D-printed components also exhibited excellent thermal stability, good mechanical properties, chemical resistance, and good imaging performance with good resolution, thus paving the way for future precision optical imaging.

Optical metasurfaces have received significant interest in the imaging community due to their ability to integrate complex optical functionalities into ultrathin, wavelength-scale devices. These engineered surfaces enable the miniaturisation of conventional optical systems, replacing bulky components with compact alternatives while maintaining high performance. These optical structures are composed of subwavelength-scale nanostructures designed to control the phase, amplitude, and polarisation of light at the nanoscale. Metasurfaces have been used to realise advanced optical elements such as lenses [92], vortex plates [41], wave plates [93], and holograms [94], demonstrating both versatility and efficiency. Nowadays, nanoscale 2D single-layer surfaces fabricated through lithographic methods severely restrict the potential use of multifunctional designs. TPP-DLW has the capability to support nanometre-scale volumetric integration, and thus the realisation of these structures has caught the attention of the research community.

Designed explicitly for focusing, metalenses are metasurfaces that enable diffraction-limited light focusing and support advanced functions, such as chromatic aberration correction, polarisation control, and multi-wavelength operation without the need for additional optical elements. One example of a metalens was given by Carmes *et al* [95], who topologically optimised 3D-printed multifunctional optical devices in the 1550 nm wavelength region. The authors experimentally demonstrated a two-layer optical concentrator fabricated in a low-index material through TPP, allowing for the focusing of incident light from five different angles at the same wavelength-scale spot.

OF imaging is performed through the collection of scattered light using bulky GRIN or ball lenses, or via TPP-DLW-printed microlenses, as mentioned above for the study described in [70] (see figure 4.7(e)) and also in [96]. However, their geometric constraints result in significant chromatic aberration that obscures images. Control over lens chromatic dispersion has already been described through the nano 3D printing of GRIN lenses (see figure 4.11). However, the realisation of such a lens on the tip of OFs has not been described, and this could limit the range of fibre-based imaging applications. Some methods to realise metasurfaces at the tip of an OF have already been demonstrated, such as ion-beam lithography [97] and chemical etching techniques [98]. However, these techniques are unable to create arbitrary structures capable of efficiently modulating light. On the other hand, EBL [99] and nanoimprinting [100] offer enough resolution for the development of these structures. However, preparing the fibre tip for planar surface patterning and predesigned pattern transfer is challenging. TPP-DLW provides a competitive alternative, allowing the direct integration of 3D diffractive elements; therefore, based on this approach, Ren *et al* [101] designed and nano-printed a 3D achromatic diffractive metalens onto the tip of a commercial SMF (see the schematic representation in figure 4.14(a)), which they designated in their study as a metafibre.

The fabricated metafibre (see the SEM image in figure 4.14(b)) exhibited achromatic and polarisation-insensitive focusing across the entire near-infrared telecommunication region [101]. While metalenses offer a new route to miniaturise fibre-optic imaging systems, chromatic aberration is always present, making them unable to operate in broadband imaging. Thus, to explicitly demonstrate the advanced performance of the proposed achromatic metafibre for broadband imaging, the authors provided detailed numerical and experimental imaging results for achromatic and chromatic metalenses implemented in SMFs. Experimental confocal imaging tests were performed using a 1951 USAF resolution test chart (see the procedure in figure 4.14(c)), and the corresponding imaging results shown in figure 4.14(d) revealed the enhanced performance of the achromatic metalens, demonstrating its ability to produce focused, sharp images with a spatial resolution of 4.92 µm under broadband light emission. In addition to the good on-axis imaging, the metalens also demonstrated good performance for off-axis imaging at angles of up to ±7.5°, corresponding to a FOV of 104 µm in the focal plane. A representative schematic of the achromatic metafibre device fabricated in this study is shown in

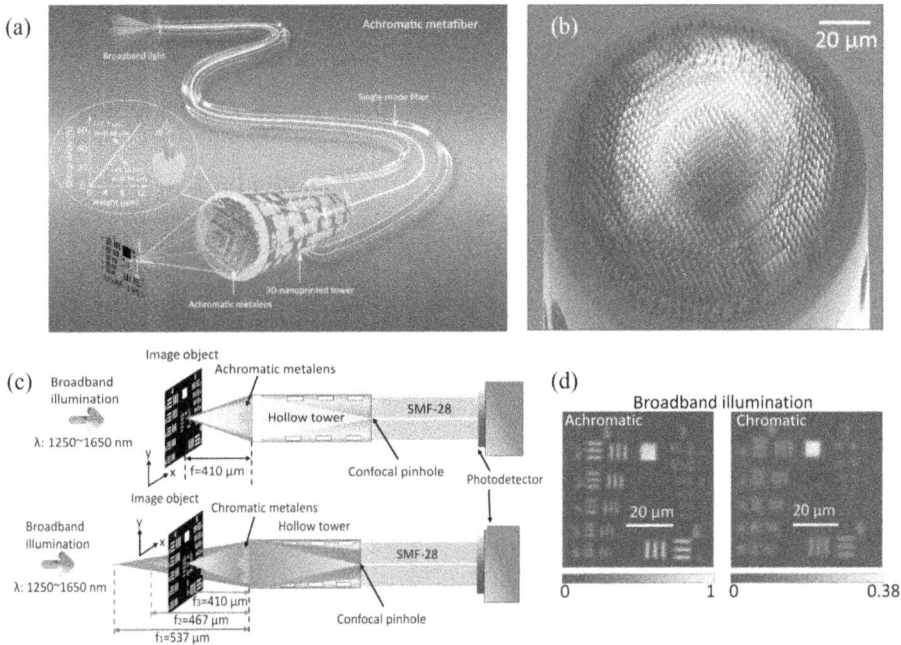

Figure 4.14. (a) Representative schematic of the achromatic metafibre structure used for applications such as focusing or imaging. (b) SEM image of the tip of an SMF showing the 3D achromatic metalens. (c) Representation of the experimental confocal tests carried out using the USAF resolution test chart for achromatic (top) and chromatic (bottom) metafibres under broadband illumination (i.e. eight equally spaced wavelengths from 1250 to 1650 nm). (d) Imaging results for the achromatic (left) and chromatic (right) metalenses. Group 6 Element 5 was used to determine the spatial resolution of the achromatic metafibre, which was 4.92 μm. Reproduced from [101]. CC BY 4.0.

figure 4.14 [101]. Through this study, the authors unleashed the full potential of fibre meta-optics for widespread applications including hyperspectral endoscopic imaging, fs laser-assisted treatment, deep tissue imaging, wavelength-multiplexing fibre-optic communications, fibre sensing, and fibre lasers [101].

4.4 Waveguides and devices at the facets of optical fibres

Nowadays, waveguides such as OFs have reached a plateau in the number of their functionalities. Therefore, to continue meeting the high demand for new high-performance OF devices capable of sustaining continuous market growth, it is necessary to develop and implement innovative new strategies. Currently, some of the most reported methods for the manufacture of micro-optical devices are subtractive methods. The subtractive methods consist of three main groups of technologies, namely laser ablation [102], polishing [103], and chemical processes [104]. Although these technologies are used to manufacture micro-optics, they tend to create structures with high surface roughness and low stiffness, resulting in waveguides with high scattering losses and fragility, respectively. Other conventional technologies that can also integrate microstructures at the end face of an OF include laser micromachining [105], focused ion-beam milling [106], and nanoimprinting [100]. These technologies face limitations in generating intricate geometries and in resolution. TPP, on the other hand, can manufacture microstructures at subwavelength resolutions, with high precision and a high degree of freedom on different substrates, such as the tips of OFs. This ability to manufacture microstructures was demonstrated in [96], where an axicon lens was manufactured with high resolution at the flat terminal of an SMF. Another example of this ability was demonstrated in [107], where collimated microlenses were 3D printed with high surface smoothness and good transparency.

The development of TPP technology also led to the development of other freeform structures on the tips of OFs, such as chiral photonic crystals able to control polarisation [108]. Another example of complex structures printed on the tips of OFs is the DLW of diffractive phase plates with sufficient quality to allow the intensity of the light that emerges from the SMF to be spatially distributed [109].

TPP has also been investigated for applications in spectroscopy, particularly in the field of surface-enhanced Raman scattering (SERS), which is widely used for chemical and biological sensing. An interesting study in this field involved the integration of TPP-fabricated SERS structures directly onto the tip of an OF, contributing to the advancement of lab-on-fibre technologies [110]. The fabricated structure was designed to detect analytes in a liquid environment and consisted of a SERS-active region (inner region) and a parabolic mirror (outer region), both positioned at the end of a multimode OF, as detailed in the schematic of figure 4.15(a).

In the schematic of figure 4.15(a), the excitation laser light exiting the OF is collected by the parabolic mirror and refocused onto the surface of the SERS body to excite SERS signals from analyte molecules in contact with the SERS-active nanostructures. The liquid analyte molecules can enter the active area through the openings at the top and side of the dome-like mirror. The Raman-scattered light from the SERS body is then collimated by the parabolic mirror and sent back through the multimode OF for detection. The SERS body comprises a half-spherical surface wrinkled periodically along its latitude to produce nanogratings (see

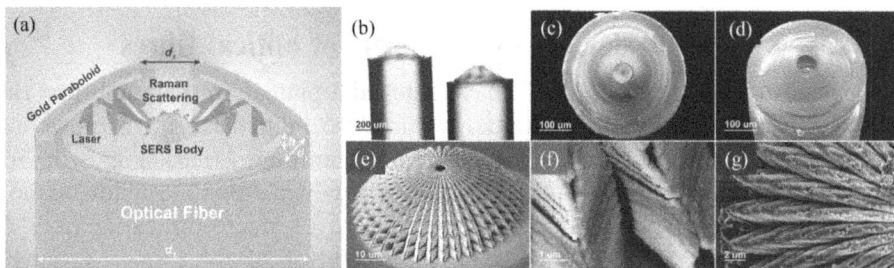

Figure 4.15. (a) Schematic of the operating principle of the SERS sensor. (b) Image of the photoresist mould for the SERS sensor before the metallisation process. (c) and (d) Top and side images of the SERS sensor after the metallisation process. (e) SEM image of the whole SERS sensor. (f) and (g) SEM images of the nanostructures of the SERS sensor [110] John Wiley & Sons. © 2015 WILEY-VCH Verlag GmbH & Co. KGaA, Weinheim.

figures 4.15(d) and (f)), which ensure that the surface plasmon resonances (SPRs) are efficiently excited by the laser light, and further wrinkled along their longitude to produce microridges, increasing the effective contact area of the SERS body with the analytes [110]. To produce these intricate microdetails (the period of the nanograting needed for SPR excitation is 590 nm), the authors took the opportunity offered by TPP-DLW printing. The following steps consisted of the thermal evaporation of Au, followed by SiO_2, and subsequently a second Au layer that was converted to nanoparticles through UV processing. This final process is required to enable the SERS activity of the SERS body, while the dielectric SiO_2 layer is, according to the research literature, used to enhance the signal. Finally, for the characterisation, crystal violet dissolved in ethanol was chosen as an analyte. The results showed a detection limit of 10^{-6} mol l^{-1}, which is lower than the detection limit of the best-performing SERS fibre sensors based on silver (i.e. sensors known to have a higher SERS effect) but with the advantage of being less prone to oxidation [110].

With the development of AM technology at the micro- and nanoscale, new fields have been explored to manufacture waveguides at small scales. Photonic integrated circuits (PICs) are one common example in the literature. In such systems, the hybrid integration of multiple optical components is required. Thus, the development of efficient coupling architectures has already been explored, as discussed in chapter 3. However, in simple terms, for efficient coupling, the mode sizes of the devices being coupled must be matched. This has been accomplished through different solutions, such as tapering structures [111], lensed OFs [112], tapered OFs [113], or the use of lenses between the OF and the chip [114]. Other elegant approaches have also been explored through the DLW of microlenses at the chip and fibre facets [88], allowing for the reduction of the transverse and longitudinal alignment complexity between the two (see section 3.3 of chapter 3). However, these solutions mostly rely on the edge-coupling nature, which limits the number of input/output ports on a chip. Thus, out-of-plane coupling from the top of the chip has been proposed. Grating couplers [115] are potential candidates. However, they are wavelength- and polar-isation-dependent, limiting their broadband operation. One interesting approach reported in the literature has been the DLW of a near-adiabatic mode converter [116]. For this, the authors directly 3D printed an adiabatic transition polymer up-taper on top of a silicon nitride chip that already included a single-mode down-taper waveguide. This allowed for the coupling of the fundamental guided mode between the PIC and the 3D-printed polymer waveguide. The polymer waveguide was then lifted from the surface of the chip through a smooth transition curvature, which reduced losses. The rectangular cross-section of the polymer waveguide was then converted to a circular cross-section to mimic the circular symmetry of the fibre. Finally, a spherical lens was added to the front of the polymer waveguide to focus the outgoing beam. A representative schematic of the proposed optical coupler is shown in figure 4.16(a), while its 3D-printed version is shown in figure 4.16(c).

The results presented in figure 4.16(c) confirm the successful direct 3D printing of a contact-free polymer structure, capable of performing optical coupling from standard SMFs on top of the chip to planar photonic circuitry. The authors tested the coupling efficiency of this approach, as shown in figure 4.16(b), measuring low

Figure 4.16. (a) Schematic of the polymer coupler (in blue) attached to silicon nitride waveguides (in orange). The two straight extended parts are used for structural support. (b) Hybrid photonic circuit utilised for the transmission tests (the silicon nitride waveguide connects both of the polymer couplers). (c) SEM image of the TPP-DLW polymer coupler. Reproduced from [116]. CC BY 4.0.

coupling loss (i.e., 1.8 dB) with a flat response over a broad wavelength region (between 1480 and 1620 nm). Additionally, the 3D couplers allowed for relaxed mechanical alignment with respect to OFs, achieving a 1 dB alignment tolerance of 5 μm in the x- and y-directions and 34 μm in the z-direction [116].

The light coupling success achieved in [116] strongly depends on the mode-matching efficiency achieved at the taper transition. In connection with this, different TPP-DLW taper geometries have already been tested [118, 119]. However, the design versatility of TPP-DLW allows for the production of tapers beyond basic geometrical shapes. Examples include the fabrication of tapered waveguides based on photonic crystal fibre (PCF) designs [119]. PCFs are single-material OFs in which an array of microscopic air holes runs along the fibre length to allow light guidance in a manner similar to the total internal reflection (TIR) mechanism. The geometric design of the air-hole channels in PCF allows for easy control and tuning of the fibre waveguide parameters, such as optical mode size and shape, modal dispersion, birefringence, and nonlinearity [120]. Currently, the most common method for producing a PCF is through the use of a fibre drawing tower. However, the PCF air-hole structure has limitations in terms of structural design freedom [121, 122]. The additive manufacture of fibre preforms provides flexibility in fibre design, as demonstrated in [123, 124] and already described in chapter 3. However, during the fibre drawing process, it is necessary to maintain the same hole diameter and hole-to-hole spacing, which represent additional parameters to be controlled. Consequently, the produced fibres exhibited significant inhomogeneities along their length. More recently, thanks to the micron–nanosize printing capabilities offered by TPP-DLW, along with its design freedom, the direct fabrication of various PCF designs on top of SMFs has been demonstrated [125]. Regarding the use of these PCF designs as mode field converters/adaptors, the authors of [119] described the direct 3D printing of up- and down-tapers on top of conventional SMFs. SEM images of these compact structures are shown in figure 4.17.

For the up-taper structure, the mode field diameter expansion ratio was 1.7, and for the down-taper structure, the mode field diameter reduction ratio was 3. The performance of the tapers was comparable to or even better than that of their competitor's step-index tapers [117, 118], while providing greater robustness due to the air hole cladding structure, which eliminated the need for additional cladding

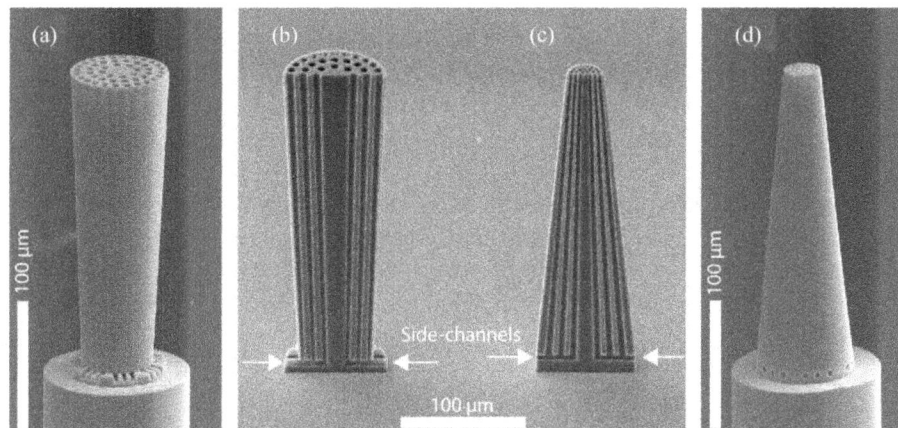

Figure 4.17. SEM images of: (a), (b) up-taper and (c), (d) down-taper structures created on top of standard SMFs. (b) and (c) correspond to cross-sectional views of the tapered structures. The side channels at the bottom were intentionally added to the development process. Reprinted with permission from [119]. © 2020 Optical Society of America.

material (commonly used in step-index waveguides to protect them from the environment) and made it easier to control the air filling fraction necessary to achieve single-mode behaviour. In addition to all these features, the PCF-like structures can also incorporate other characteristics, such as birefringence, allowing for the inclusion of additional features.

4.5 Reflective optics

The advancement of modern photonic systems has spurred significant interest in the miniaturisation and customisation of optical components. Among these, reflective optics at the micro- and nanoscale have played an important role in the design of such structures. TPP-DLW technologies allow the fabrication of complex 3D geometries with submicron precision, making them ideally suited for producing micro-mirrors, parabolic reflectors, and beam-steering structures directly on optical platforms.

Typically, reflective micro-optics produced via TPP-DLW can involve one- or two-step procedures. The most straightforward method uses the TIR condition, where the geometry of the structure to be fabricated is designed to reflect light within the polymerised photoresist structure, avoiding the use of further processing. Conversely, for a two-step procedure, the TPP-fabricated structure can also be engineered to include a highly reflective surface added to the newly formed TPP structure through a metallic coating (e.g. gold or aluminium) deposition process.

One of the first reflective optics realised through AM was proposed by Malinauskas *et al* [126], in which a 15 × 15 μm 45° prism was manufactured on a coverslip substrate by DLW using a germanium–silicon photosensitive hybrid material. The surface quality of the prism was smooth without defects, and this was assumed to be related not only to the hatching distance but also to the properties of the composite itself. The authors also assumed that, under these conditions, micro-optical elements could be printed with $\lambda/20$ precision for both visible and infrared regions. Another simple reflective structure was also proposed by Bianchi *et al* [127], who reported the TPP fabrication of a simple parabolic reflective structure on the end surface of a multimode OF with a 50 μm core. The reflective structure was designed to enhance the numerical aperture of the OF (i.e. from 0.22 to 0.93), allowing it to reach high resolutions. In addition to reaching extremely high numerical apertures, the parabolic reflector geometry did not rely on refraction and, therefore, featured a focusing power that was independent of the immersion medium. Through this study, the authors showcased the potential of TPP to create complex reflective structures directly on OF facets, paving the way for advanced applications in endoscopy and optical manipulation.

Probing photonic integrated circuits at the wafer level is essential for ensuring reliable process control and effective performance evaluation in cutting-edge production workflows. Thus, the TPP-DLW of reflective components has recently been utilised to demonstrate free-space coupling between different optical wave-guides [88]. For this, a free-form mirror with curved surfaces was designed to simultaneously adapt the mode profile and the propagation direction of light. The structure comprised a 3D-printed TIR free-form mirror, designed to simultaneously focus and redirect light in the desired direction. The viability of these TIR mirrors was first tested using two orthogonal SMF terminals, as indicated by the schematic shown in figure 4.18(a), and the TIR mirror in the SEM image shown in figure 4.18 (b). The results revealed low coupling losses (0.6 dB), corresponding to an efficiency of $\eta = 87\%$. The successful results led the authors to implement TIR mirrors in the coupling between the fibre and the chip, as indicated in the schematics of figures 4.18 (c) and (e). This led to a loss of 2.9 dB ($\eta = 51\%$) for a free-form TIR mirror directly printed on the facet of an edge-emitting indium phosphide (InP) laser, as shown in the SEM image in figure 4.18(d). A loss of 1.1 dB ($\eta = 51\%$) was measured when coupling a vertical-cavity surface-emitting laser (VCSEL) with an OF containing the 3D-printed mirror, as shown in figure 4.18(f). The results demonstrated the viability of coupling low-cost surface-coupled lasers or photodetectors to SMF arrays while maintaining compact and flat form factors. Additionally, the results demonstrated

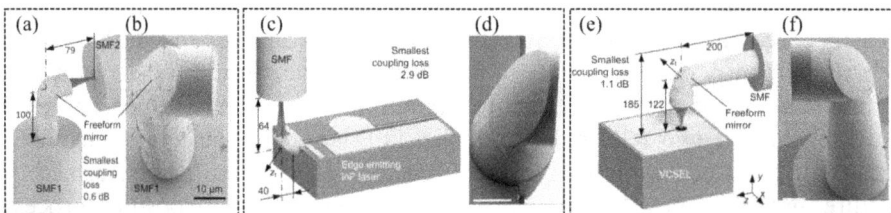

Figure 4.18. Representative schematic of the coupling of light between different optical components using DLW TIR mirrors on the facets of optical components. Light coupling between: (a) two SMFs; (b) an InP laser and an SMF; (c) a VCSEL and an SMF. SEM images of TIR mirrors on the facets of: (b) an SMF, (d) an InP laser, (f) an SMF. The Z_t axis is tilted by 45° relative to the optical axis. Adapted from [88], with permission from Springer Nature.

the possibility of replacing the angled polished fibre arrays and complex micro-optical assemblies, including ball lenses and microprisms, commonly found in commercial silicon photonics (SiP) products.

The optical probing of surface-coupled devices, as described above, yielded acceptable results. However, wafer-level probing of edge-emitting devices with hard-to-access vertical facets at the sidewalls of deep-etched dicing trenches was still unexplored. Taking this into account, the same team joined efforts to address this challenge using TPP-DLW of free-form coupling elements on the end faces of SMFs, which fitted into deep-etched dicing trenches on the wafer surface [128]. The DLW elements comprised TIR mirrors for redirecting the light from an out-of-plane to an in-plane direction, followed by an aspheric lens used to match the mode field to that of the waveguide to be coupled. The approach was tested using edge-emitting waveguides on various integration platforms, including SiP, silicon nitride (TriPleX), and InP, achieving coupling losses as low as 1.9 dB and demonstrating a reliable method for testing wafer-level edge-coupled PICs.

Another interesting method for achieving out-of-plane coupling in integrated photonics is through the use of prism couplers. These prisms utilise a phase-matching mechanism that aligns the projection of the incident light wave vector with the waveguide mode. They are known for their theoretical 100% directionality and coupling efficiency, making them extremely interesting. The most common configuration for prism couplers is the Kretschmann configuration [129]. This configuration requires direct contact between the prism and the substrate surface, meaning that the material of the photonic chip must be transparent at the operational wavelength. To avoid this problem, another configuration has been proposed, namely the Otto configuration [130]. In this case, the prism coupler is separated from the chip by a distance similar to the wavelength. However, this leads to another problem related to the precise control of the gap thickness. This led Safronov *et al* [131] to propose the use of TPL for the direct writing of a microprism, based on the Otto configuration, onto the surface of a photonic crystal, allowing the coupling of incident focused beams normal to the chip surface, as indicated in the schematics of figure 4.19.

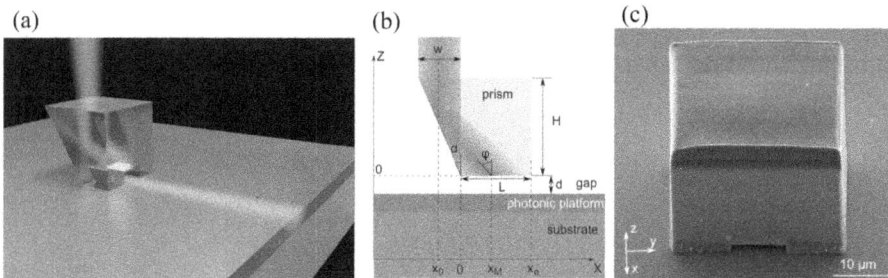

Figure 4.19. (a) Rendered view of the out-of-plane microprism coupler in action. (b) Two-dimensional schematic of the microprism with all the parameters needed for optimisation. (c) SEM image of the microprism [131] John Wiley & Sons. © 2022 Wiley-VCH GmbH.

As can be seen in figures 4.19(a) and (b), the light beam incident on the prism and normal to the surface of the optical chip is reflected at the slanted face of the prism in the direction of the gap between the prism base and the waveguide surface of the optical chip. The surface electromagnetic wave is excited under the prism base due to the phase-matching condition caused by frustrated TIR. At the end of the gap, the waveguide mode beneath the prism is transformed into a waveguide mode that propagates freely along the platform's surface or within the waveguide on it, thus enabling highly efficient unidirectional coupling. Through the optimisation of the parameters shown in figure 4.19(b), the authors reported theoretical coupling efficiencies close to 80%. Experimentally, the authors validated the concept on a Bloch surface wave (BSW) platform by coupling a 10 µm Gaussian beam to the BSW using a TPP-fabricated microprism (see the SEM image in figure 4.19 (c)), which had a size of 30 µm and a gap of about 300 nm [131]. The results revealed BSW excitation efficiencies of more than 40%, which were better than those reported in [132] using a grating for BSW excitation.

Apart from focusing and redirecting light, reflective optics have also advanced into fields such as biology, where they are used for cell trapping with optical tweezers. Optical trapping typically involves the use of complex systems, such as bulky microscopes [133, 134] or counter-propagating beams [135, 136]. On the other hand, methods for on-chip trapping, such as planar photonic chip trapping [137], optical resonators [138], and plasmonic structures [139], have employed the evanescent field to perform optical manipulation. Because the evanescent field only extends over nanometre-scale distances, optical trapping is restricted to two dimensions near the chip surface. This limitation triggered the investigation of new techniques for enabling three-dimensional optical trapping, leading to the development of optical reflectors directly integrated into a microchip [140]. These optical reflectors were designed to converge light from two waveguides to a single point. Figure 4.20 shows schematics of these micromanipulators [140].

The proposed design enabled optical trapping in a 3D space, specifically the space above the chip, addressing the limitations of the 2D trapping methods used by other technologies. Another notable feature of this optical tweezer was its excellent optical efficiency (>95%), which enabled a low trapping power threshold. In [140], optical trapping power values of 19, 17, and 30 µW were observed for the x, y, and z dimensions, respectively, which are notably lower than the values found in the existing literature on other optical tweezers [141, 142].

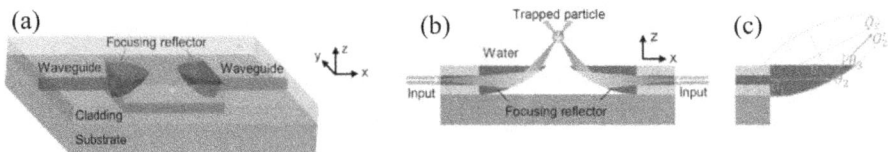

Figure 4.20. (a) Three-dimensional view and (b) 2D cross-sectional view of the on-chip free-form focusing reflectors used for particle trapping, as illustrated in (b). Part (c) is a close-up of the free-form focusing reflectors shown in (b). Adapted with permission from [140]. © 2021 Optical Society of America.

As described earlier, most reflective optics studies are based on the TIR principle. However, in certain situations, such as when the external refractive index differs from that of air, the use of reflective optics can be more challenging and may even become impractical due to the limitations of TIR. This creates the need for surface coatings to enable those surfaces to be effectively reflective. Additionally, changes in the external refractive index can lead to variations in the reflection angle, making reflection problematic in certain applications where the refractive index is prone to variation. Optical tweezers may benefit from this coating process. The proof of concept for this was demonstrated in [143], which reported the fabrication of two face-to-face microprisms containing a metallic surface on the slanted faces [143]. This involved the fabrication of the microprisms using the TPP method, followed by the deposition of thin gold on the entire structure. To ensure that the gold was present only on the slanted parts of the prisms, a protective layer of photoresist was applied only to the slanted surfaces of the microprisms, and an etching process was subsequently executed to remove the gold from the unprotected areas. Following this step, the structure was finally able to function as an optical trap. A representative schematic of the prism manufacturing process is illustrated in figure 4.21 [143].

The small size of OFs makes them an attractive alternative for implementing optical tweezers. However, to make them functional, special fibre post-processing techniques involving tapering have been routinely implemented. The advent of TPP-DLW enabled the easy integration of micro-optics at the tips of OFs. Considering this opportunity, Liberale et al [144] reported the DLW fabrication of microprism reflectors at the end face of an OF bundle composed of four OFs. The proposed optical tweezer allowed on-chip manipulation, Raman spectroscopy, and fluorescence microscopy of individual cells. This innovative technology also demonstrated the potential for integration into a microfluidic circuit, enabling the development of a lab-on-a-chip.

The reflective optical components described in the previous paragraphs rely on the TIR condition. While TIR-based designs are highly effective, certain applications may involve incident angles that do not meet the TIR condition, thereby limiting their use. To address these limitations, an alternative class of reflectors—metal-coated reflectors—has been explored. In such designs, a thin metallic film is deposited onto the surface of the printed structure, typically through vapour deposition techniques. These coatings enable broadband, angle-independent reflection due to the high reflectivity of metals, making them suitable for scenarios where TIR cannot be achieved. One example of such reflectors has been presented by Atwater et al [145], who reported the TPP fabrication of a microphotonic parabolic light director. These devices use the focusing properties of compound parabolic concentrators and, if combined in array formats, they could find opportunities in optoelectronic applications, namely for the collimation of light emitted by light-emitting diodes and also to improve the efficiency of solar cells through the collimation of the collected light to reduce the entropy changes associated with light emission. The authors of [145] reported the TPP fabrication of microphotonic parabolic light directors on top of a coverslip, with a height of 22 μm and a diameter

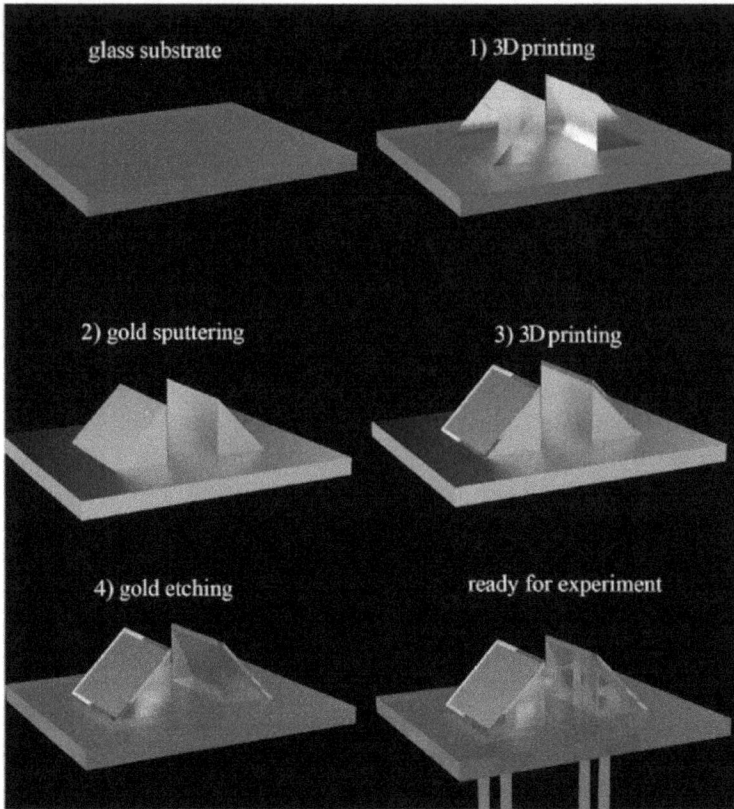

Figure 4.21. Representative schematic of the fabrication process of the metalised microprism mirrors. Reproduced with permission from [143]. © (2020) COPYRIGHT Society of Photo-Optical Instrumentation Engineers (SPIE).

of 10 µm. These were later coated with 20 nm of chromium and 380 nm of silver through plasma sputter coating, followed by the addition of apertures at the bottoms of the paraboloids through FIB etching. Figures 4.22(a) and (b) show SEM images of the silver paraboloid light directors, both arranged in an array and as an individual unit, respectively [145].

The transmission characteristics of the fabricated parabolic light reflectors demonstrated strong beam directionality with a maximum divergence of 5.6 °, indicating their potential use in generating collimated beams for applications in advanced solar cell and light-emitting diode designs [145].

The manufacture of coated surfaces has also been explored for other applications, such as in the sensing field, namely for the development of a monolithic hinge mirror at the tip of an OF, which resembled a Fabry–Pérot (FP) cavity resonator and could be used as a temperature sensor [146]. The hinged structure was printed using a TPP system. After this, the curved structure inside the printed device and the surface of the OF were coated with a layer of gold using a plasmon sputtering machine. The coated structure had a temperature sensitivity of 190 pm °C^{-1}, showing better

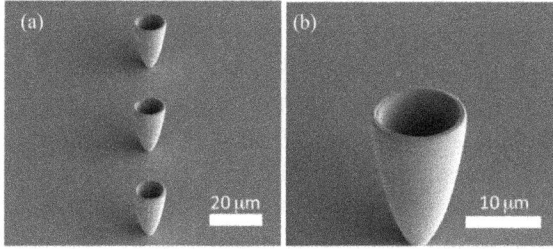

Figure 4.22. SEM images of: (a) an array and (b) single parabolic reflectors coated with silver. Reproduced with permission from [145].

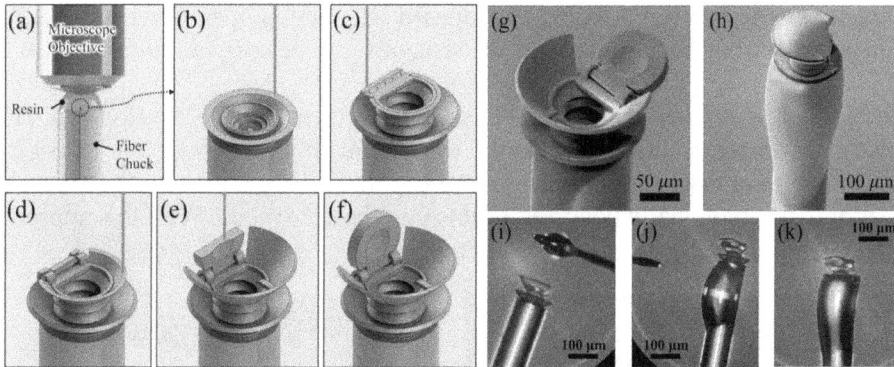

Figure 4.23. (a–e) Step-by-step views of the DLW of the hinge structure. (g–h) Coloured SEM image of the hinge structure. (i–k) Microscope images of the hinge during the sealing process of the gold-coated top cover. Reproduced from [148]. CC BY 4.0.

sensitivity than a structure without the coating [147]. A representative example of the hinged structure is shown in figure 4.23.

4.6 Polarisation control optical elements

In modern optics, the polarisation of light is a crucial parameter that requires precise control. Its significance stems from its wide range of applications, including enhancing contrast in photography and boosting the efficiency of optical communication systems. Consequently, substantial research has been dedicated to developing methods for manipulating polarisation. Traditional optical devices typically rely on the intrinsic properties of materials to achieve polarisation control. However, 3D-printed optical components offer a different approach. In these components, polarisation manipulation is not solely dependent on material properties. Instead, it primarily arises from carefully engineered structural features that exploit optical phenomena such as refraction, reflection, and diffraction.

Currently, several free-space bulk optical devices have been replaced by miniaturised components that perform similarly on a micrometre scale. OFs hold the potential to replace most of the existing free-space optical components. Different micro- and nanotechnologies have been used to allow OFs to manipulate light at small scales [149]. EBL and FIB milling are technologies capable of achieving unprecedentedly high resolutions, and they have been used to fabricate various micro- and nanostructures at the tips of OFs. One example of this type of device was a polarisation diffraction grating produced through FIB milling at the end surface of an OF [150]. With a lower resolution but with much more freedom, TPP-DLW offers a reliable solution for developing microdevices at the tips of OFs. Due to the interest in polarisation devices and considering the opportunities of TPP, the DLW of a polarisation beam splitter (PBS) has already been reported in the literature. One of the earliest studies was reported by Hann *et al* [151], who manufactured a PBS on top of a standard SMF. The device consisted of a refractive prism that was able to redirect the incident light to a subwavelength grating, which separated the two polarisations. A rendered view of the proposed structure in operation can be seen in figure 4.24(d), while SEM images of the structure can be seen in figures 4.24(a)–(c) [151].

When light is injected from the fibre side, the integrated PBS developed in [151] works as a simple polariser. However, when light impinges from the air side, it allows for further applications, such as the capability to simultaneously couple the power of two orthogonally polarised beams into the fibre mode. With this approach,

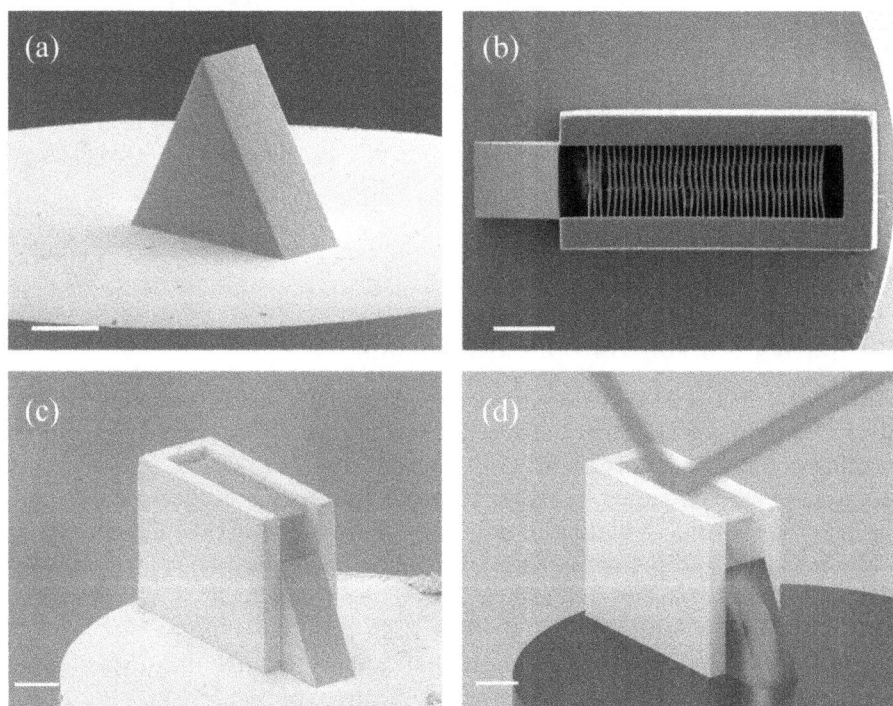

Figure 4.24. (a) Coloured SEM image of the TIR prism. (b) Top view SEM image of the PBS. (c) Coloured SEM image of the PBS composed of the prism (blue), the grating (orange), and the support structure (green). (d) Rendered view of the PBS in action. The scale bars are 10 μm. Reprinted with permission from [151]. © 2018 Optical Society of America.

no power is lost (apart from reflection losses), resulting in superior performance compared to a non-polarising beam splitter. Characterisation of the fabricated PBS at a wavelength of 1550 nm showed a polarisation purity superior to 80% for the emerging transverse magnetic (TM) and transverse electric (TE) polarisation beams.

Another example where AM contributed to the manufacture of polarisation control elements was given by Wei *et al* [152]. Their study consisted of the manufacture of a compact free-space free-form inverse-designed PBS for the infrared region, able to manipulate light propagating parallel to the substrate in a photonic chip device. The proposed PBS was designed using an inverse-design algorithm and manufactured through DLW. The PBS acted as a meta/grating, able to split the parallel and perpendicular polarisations into left and right first diffraction orders. The transmission losses were relatively low, reaching 1.5 dB at 1550 nm and 2 dB at 1300 nm, while the experimentally obtained extinction ratio was 5 at the optimal wavelength.

Short sections of PCF-like structures have already been additively manufactured at the tips of SMFs, allowing the control of fibre waveguide parameters, such as the optical mode size, as described in section 4.4 and seen in figure 4.17. However, other interesting capabilities can also be realised with PCF-like structures. One potential field in which the PCFs can contribute is the control of polarisation by creating designs in which PCFs exhibit birefringence. This has already been described in several studies that explored the fabrication of fibre polarisers through the use of single-core highly birefringent PCFs. Conversely, for PBS, despite being theoretically proposed [153], the inherent complexity associated with the fabrication of asymmetric dual-core geometries with different hole sizes along the length of the fibres makes its practical development very challenging to achieve or even impossible through commonly used PCF drawing techniques. TPP-DLW is well suited to handle complex designs at the micro- and nanoscales, and the development of a PBS based on TPP-fabricated PCF-like structures has already been successfully demonstrated by Bertoncini *et al* [125]. In doing so, the authors relied on the concepts developed in [153]. The PBS consisted of a PCF-like structure created on top of an SMF and was composed of three sequential waveguiding segments: a PCF-like tapered coupler (down taper) right after the flat surface of the SMF, followed by a dual-core directional coupler, a birefringent PCF structure, and, finally, a fan-out section able to separate the distance between the two cores. A schematic of the structure, its cross-sections in the different segment regions, and the corresponding SEM images are shown in figure 4.25 [125].

The first PCF-like section of the fibre PBS shown in figure 4.25(a), i.e. the magenta-coloured region in the schematic, is composed of a down taper that couples the 6 μm mode field diameter of the SMF to one of the two cores of the dual-core directional coupler PCF (see figure 4.25(b), the blue-coloured region in the schematic), composed of an intentionally nonsymmetric structure with dimensions of 1×2 μm and a core-to-core separation of 2.4 μm, allowing for a short coupling length, namely 140 μm for the optimised design structure. For the last waveguide segment of the PBS (seen in figure 4.25(c)), the two orthogonally polarised beams travelling in each fibre core underwent an adiabatic transformation from an asymmetric mode shape profile to a circular mode shape. The distance between

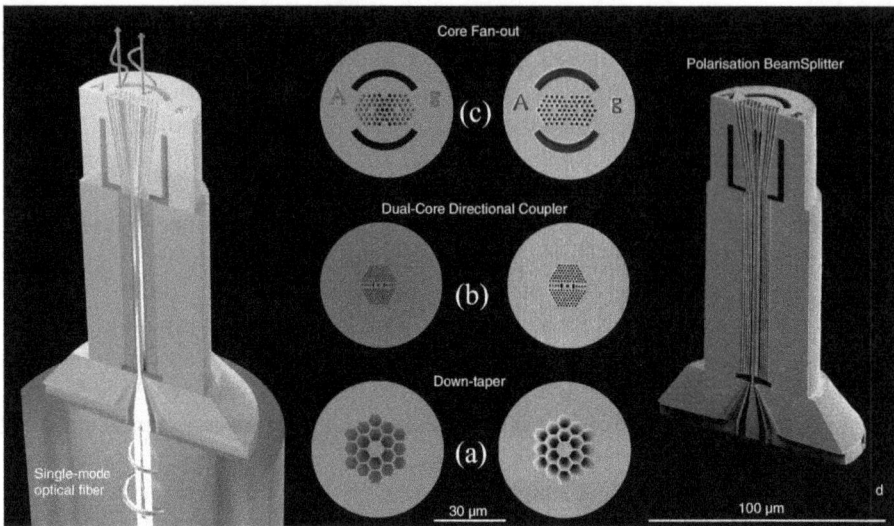

Figure 4.25. (left) Representative schematic of the PCF-based PBS on top of an SMF. The coloured cross-sectional images demonstrate its different sections: the down-taper (magenta); the directional dual coupler (blue), and the core fan-out (cyan). The representative schematic also shows the input central beam (yellow) with an arbitrary polarisation that is split into its horizontally (red) and vertically (green) polarised components. (right) SEM cross-sectional images of the 3D-printed PCF structures. Reprinted with permission from [125]. © 2020 Optical Society of America.

the cores was also increased to 10 μm to easily distinguish the cores during the characterisation tests [125]. This last section can be adapted to specific coupling scenarios; for instance, PICs with specific mode field specifications and waveguide separations. Considering all three segments, the structure was about 210 μm tall. Regarding the splitting performance, the two cores of the PCF PBS showed extinction ratios that simultaneously exceeded 10 dB over a 100 nm bandwidth centred at around 1550 nm. At 1550 nm, the authors reported a minimum extinction ratio of 14.4 dB [125]. While demonstrating a PBS at the tip of an OF, this study also showed the powerful tool TPP can be for the manipulation of light at the micro- and nanoscale. While the prior paragraphs were dedicated to diffraction-based PBS, which relies on free-space output beams, this approach can easily control the beam shape and location of guided beams, making it more attractive.

Regarding the insertion losses of the PBS described in [125], the authors reported values of 1.35 and 1.18 dB for the vertical and horizontal polarisations, attributing these to the short length of the down-taper used. However, they also attributed those losses to the polymer material losses, stating that the use of glass ceramics could be a potential solution to reduce system losses. The problem with polymers is not only due to their poor transparency. Polymers can also easily deform at low temperatures, have poor mechanical stability, and can even absorb water from the environment, making them undesirable for most precision optics applications. Based on these poor characteristics, several studies have already reported the micro 3D printing of inorganic glasses, typically involving the use of sintering processes at high temperatures, i.e. 1300 °C [154], and low temperatures, i.e. 600 °C [91, 92], or without thermal treatment through the use of HSQ selectively crosslinked into silica glass through nonlinear absorption when exposed to 1040 nm laser pulses [89], as discussed earlier in section 4.3. Sintering processes, even at low temperatures, restrict the host materials that can be used. This prevents the use of OFs, since they include a polymer coating and jacket that can be damaged by temperature. Thus, the HSQ method appears to be a suitable candidate for implementing submicron structures in OFs. Because of this, [155] reported the development of an ultracompact PBS at the tip of an OF. The structure was composed of two subwavelength grating blocks with orthogonal grating orientations. Each block had a period of 1.1 μm, while the thickness of each nanoplate was 0.9 μm, and its height was 4.5 μm. The structure was tested in the 1550 nm region by injecting light with different polarisation orientations; the recorded output beam profiles confirmed its capability to perform polarisation control and beam steering.

In addition to polarisers, the control of polarisation is commonly implemented through other bulky free-space optical components, such as polarisation rotators and phase retarders. The phase retarder, also known as a wave plate, is an optical device that delays one component of the electric field of light relative to another, thereby altering the state of polarisation of the transmitted signal. This occurs because one component travels faster than the other, introducing a phase shift between them. Among the different phase retarders, the quarter-wave plate (QWP) and the half-wave plate (HWP) are the most popular. Considering the miniaturisation capabilities of DLW and taking into account the straightforward integration of printed components into substrates and optical elements, Bertoncini et al took the opportunity to fabricate a Fresnel rhomb at the tip of a polarisation-maintaining (PM) OF [156]. A Fresnel rhomb works as a QWP and consists of a rhombohedral prism. Light entering it undergoes two total internal reflections at a precisely designed angle of incidence, α (see figure 4.26(a)), so that the total phase shift difference between the horizontal and vertical polarisation components reaches 90° [157]. In contrast to phase shifts induced by material birefringence, a Fresnel rhomb operates over a wide range of spectral bandwidths. The miniaturised Fresnel rhomb developed in [156] had a total length of 320 μm and consisted of a 45° angle between the rhomb axis and the PM fibre's axis. To ensure the proper operation of the device, the diverging beam exiting the OF was collimated through a properly designed spherical microlens at a 150 μm distance from the terminal of the PM fibre. A schematic of the designed structure is shown in figure 4.26(b) [156].

Through alignment of the input polarisation along the slow and fast axes of the PM fibre, the structure was able to generate left or right circular polarisation over a spectral bandwidth larger than 300 nm. Such a large bandwidth creates opportunities for the use of this ultracompact fibre probe for remote excitation in circular dichroism-based spectroscopies, such as circular dichroism (in the UV–Vis band) and vibrational circular dichroism (in the near-infrared–IR band), as well as for Raman optical activity spectroscopy [156].

Another important polarisation control element commonly described in the research literature is the polarisation rotator. This component can rotate the linear

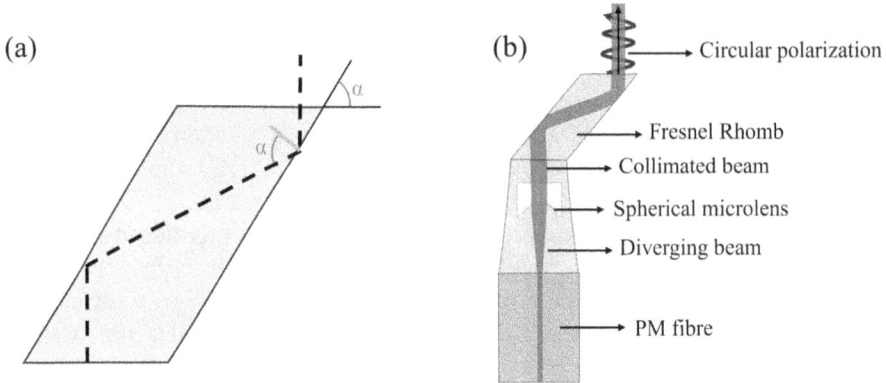

Figure 4.26. (a) Sketch of the Fresnel rhomb with a ray (dashed line) incident at a normal angle to the small side of the parallelogram. (b) Structure designed in [156], composed of two main parts: a spherical collimating microlens and a Fresnel rhomb oriented at 45° with respect to the PM fibre's optical axis.

polarisation direction and typically operates in free space. It is very common in optical communication systems, polarisation-sensitive imaging, and laser systems. On the other hand, PIC technology has made significant progress in recent years, primarily in the areas of communication, information processing, detection, and sensing. Therefore, it is necessary to develop functional components capable of processing information within a chip. In this field, polarisation-encoded information requires the use of on-chip polarisation conversion through integrated waveguides. Published reports have already demonstrated the capability to manipulate polarisation in PICs [159–161], but the technologies used present limitations, either related to wavelength sensitivity [158], the challenges of fabrication with conventional 2D lithography and dry etching techniques [159], or even poor performance due to material absorption [160]. Regarding these limitations, [161] reported the use of fs DLW for the fabrication of twisted waveguides with precisely controlled 3D features. Through adiabatic mode evolution, the twisted waveguides allowed the conversion of incident light to its orthogonal polarisation during transmission. The proposed twisted waveguides demonstrated good conversion efficiencies in both the telecom and visible wavelength regions (more than 90% at 1550 nm and 80% at 646 nm). Additionally, due to this successful achievement, the authors extended their work to the development of polarisation routers. Together with polarisation rotators, polarisation routers are essential components for polarisation-encoded information processing. These devices are responsible for selectively routing an input optical signal to target output ports with different modes through a predesigned architecture [161]. The fabrication of polarisation routers was demonstrated for a four-port router with different twisting angles, allowing photons with different polarisations to follow different paths, with low insertion losses and high efficiencies.

Still within the field of PIC technology and considering the coupling of an SMF to a highly polarisation-sensitive PIC, Nesic *et al* [162] described a proof of concept for the fabrication of a photonic wire bond between the facets of an OF and a PIC, which was achieved through the DLW fabrication of a PBS, polarisation rotators,

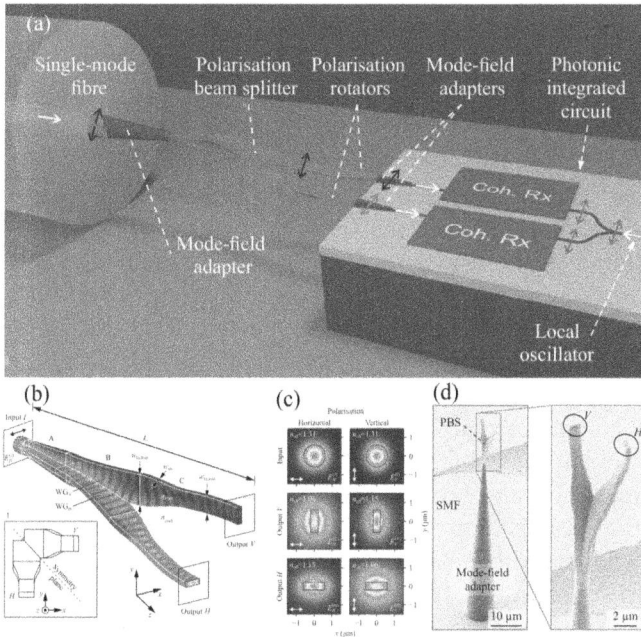

Figure 4.27. (a) Representative schematic of a photonic wire bond comprising a DLW PBS, polarisation rotators, and mode field adaptors, between an SMF with two degenerate polarisations (red and blue) and a PIC containing a dual-polarisation coherent receiver. (b) Three-dimensional model of the PBS, comprising an input waveguide port with a circular cross-section and a pair of output waveguide ports with horizontally and vertically oriented rectangular cross-sections. (c) Electric field of the fundamental modes for both polarisations at each PBS port. (d) Coloured SEM images of the PBS. The red part refers to the mode field diameter adaptor used between the SMF and the PBS input. Reproduced from [162]. CC BY 4.0.

and mode field adaptors. A representative schematic of the proposed structure can be observed in figure 4.27(a).

The proof-of-concept device shown in figure 4.27(a) links a rotationally symmetric SMF featuring degenerate polarisation states to a PIC consisting of a dual-polarisation receiver used for coherent communications. The working principle of the PBS is demonstrated in the 3D rendering of a three-port device shown in figure 4.27(b), where the input port's circular cross-section is divided into two horizontally and vertically oriented ports with high aspect ratios and rectangular cross-sections. This allows data signals from the SMF in orthogonal polarisation states to be separated and detected independently using a pair of coherent optical receivers. Within the initial segment, the circular cross-section input port is adiabatically morphed into a cross-shaped cross-section. This allows the two degenerate modes to be transformed into higher-effective-index modes polarised along the long side of the rectangle. These modes are then gradually separated in the transverse axis, allowing their coupling to other devices. The effective mode field indices at each port can be seen in figure 4.27(c). Each of the output ports can then be equipped with polarisation rotators in the form of twisted waveguides, allowing the polarisations in both output ports to rotate in the same direction. An SEM image of the PBS fabricated on an SMF by DLW can be seen in figure 4.27(d). The red region indicates the mode field adaptor and was intentionally added to match the mode field of the SMF to that of the PBS input port. The structures were later characterised, showing an 11 dB extinction ratio over a large bandwidth (between 1270 and 1620 nm). The proposed structure was later successfully demonstrated with a 640 Gbit s^{-1} dual-polarisation data signal employing 16-state quadrature amplitude modulation, achieving similar optical signal-to-noise ratio performance to that of a commercial PBS without any measurable penalties [162].

Currently, one type of material attracting attention for its potential usefulness in controlling polarisation is the liquid crystalline network (LCN). LCNs are materials composed of polymeric networks in which liquid crystalline chains are interconnected, resulting in an elastomeric material [163] that supports various phase transitions. LCNs are recognised for their distinct characteristics, including low losses, tunability, and machinability to subwavelength precision. Because of these properties, some photonic devices have already been conceptualised [164–166]. Devices related to polarisation are no exception, and the additive manufacture of a remotely controllable multichannel polarisation conversion element [167] has already been demonstrated. The proposed element consisted of an anisotropic diffraction grating manufactured through TPP. The fabrication involved local polymerisation and the cross-linking of a mixture of liquid crystal monomers. Since the grating was made of liquid crystal material, it allowed the polarisation control to be tuned, as the shape and refractive index changed when transitioning from the nematic to the paranematic phase, rendering the material more isotropic. Thus, by using light stimuli, the authors were able to control both the material and its phase, and consequently its polarisation, thereby opening this technology to a plethora of applications in polarimetry and other fields.

4.7 Fibre tip sensors

OFs have been used for sensor applications for several decades. The reason for this relates to their special characteristics, such as small size, inertness, chemical and mechanical resistance, high transparency, immunity to electromagnetic interference, and multiplexing capabilities, among others. Their success has always been tied to the growth of the telecommunications industry, which has allowed the sensor community to take full advantage of the capabilities of fibre.

Nowadays, fibre sensors have been reported to measure several parameters. The strain sensor is, by far, the most widely reported. However, this capability has also been used to indirectly measure other variables, such as pressure, acceleration, torsion, curvature, etc. In addition, fibre materials respond to temperature, allowing them to sense this parameter. If light guided by the fibre interacts with the external environment, it can also be used to measure the refractive index of the external medium, such as that of liquids or gases. Furthermore, when combined with responsive materials, fibres can measure parameters such as pH, humidity, bacteria, and viruses.

OF sensors rely on the ability to transduce an external stimulus, which can be mechanical, chemical, or biological, into a measurable change in an optical signal.

These changes are typically characterised by intensity, wavelength, phase, or polarisation. To sense such changes, fibre sensors are constructed using various fibre-optic technologies, including fibre Bragg gratings (FBGs), tilted FBGs, long-period gratings (LPGs), and a wide range of interferometers such as FP cavities, Mach–Zehnder interferometers (MZIs), Sagnac interferometers, and multimode interferometers. The fabrication of these sensors involves the use of several well-established technologies, including UV lasers for modifying the refractive index of silica and CO_2 lasers or electric arcs for heat-deforming the fibre. Such fabrication often involves specialised OFs and materials that aid or enhance the sensor's capabilities. These techniques have been exhaustively reported in recent decades, and it has become challenging to innovate further within this technology. However, to fully take advantage of the light-guiding properties of OFs, it is necessary to use technologies that can micromanipulate the waveguide at resolutions smaller than the wavelength of the light radiation. Femtosecond lasers have been widely reported for the micromachining of fibres. Additionally, FIB milling and EBL have also proven their capabilities. However, constraints related to time and cost can be problematic in fibre component development.

The subwavelength resolution of the TPP-DLW technique makes it an ideal candidate for the fabrication of micro- and nanodevices in OFs. The extreme ends of an OF make excellent substrates for the construction of these devices, since light exiting from the fibre tip can directly interact with any structure placed in front of it. This makes the fibre tip an ideal platform for realising a 'lab on a fibre'. Through the proper design and implementation of TPP structures at the tips of OFs, it is possible to explore new and exciting possibilities for fibre sensors. This is state-of-the-art technology and is certainly a technology that will shape the future of the fibre sensor field.

- **Force sensors**

Accurate force measurement at the microscale is crucial in various fields. While force sensing at the macroscale is well established, significant challenges remain when attempting to measure forces in the range of micro- to piconewtons, a range commonly encountered in fluids, air, and biological processes [168–171]. To address this gap, a recent study has explored the use of 3D-printed FP cavities at the tips of OFs for the detection of small forces at the microscale [172]. The principle behind the measurement relied on depth-sensing indentation, which is the most commonly used technique for characterising the mechanical properties of materials at the micro- and nanoscale. The sensor consisted of a suspended beam constructed at the tip of a cleaved OF. The design was intentionally created to work as an FP interferometer. Thus, it was composed of two pillar supports positioned along the same fibre diagonal and close to the fibre border terminal. They had a height of 30 μm and were connected by a thin 3 μm membrane sheet. This membrane was able to flex when pressed, and to aid this function, a probe with a 5 μm diameter and a 35 μm height was added to the central region of the membrane. The sensor was 3D printed by TPP-DLW at the far end of the fibre. A SEM image of the FP force sensor is shown in figure 4.28.

The operating principle of the sensor is that of an FP interferometer. Light travelling through the fibre reaches the core–air interface, forming the first reflection.

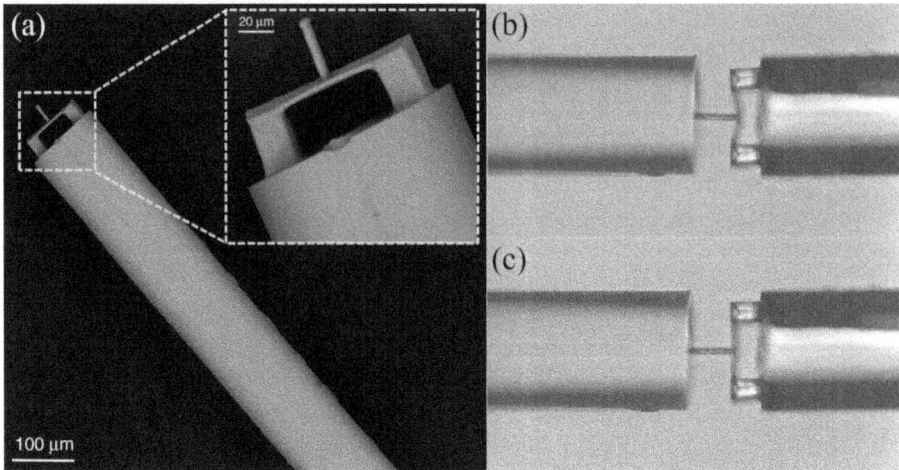

Figure 4.28. (a) SEM image of the micro-force FP sensor. Microscope images of the micro-force sensor when (b) pressed and (c) released against the tip of an OF. Reproduced from [172]. CC BY 4.0.

Then, after travelling through the air region, the remaining light encounters another reflector, this time created between the air and the 3D-printed polymer interface. When recombined in the fibre, the two reflections have different phases due to the difference in the optical paths travelled by each of the light reflections. Consequently, they interfere constructively and destructively, creating an optical spectrum with maxima and minima. The wavelength location where these maxima and minima of interference occur depends on the optical path length (OPL, which equals nL). Considering the sensor is in air ($n = 1$), any displacement of the membrane will decrease the cavity length (L), resulting in maxima and minima of interference at different wavelength locations. Through proper calibration of the observed wavelength shift for different applied forces, it is possible to establish a relationship that can later be used to measure the force by monitoring the spectral deviation of the wavelength. This was exactly what was done in [172], where the sensor was calibrated with forces ranging between 0 and 2700 nN in steps of 300 nN. This allowed the authors to achieve a force sensitivity of -1.5 nm μN^{-1}, which was one of the highest when compared to other established research [173–178]. Finally, to demonstrate the applicability of the sensor, the team used the calibrated sensor to measure the Young's modulus of polydimethylsiloxane (PDMS), a feeler of a butterfly (Danaidae), and human hair, achieving accurate results compared to those measured by atomic force microscopy (AFM).

The sensor structure reported in [172] was not optimised to allow high membrane deflections. In fact, the use of a cantilever approach would have been more suitable, since it allows for greater membrane flexibility, thereby enabling higher resolutions. Furthermore, the deflection method can only be applied to objects with regular shapes, thereby restricting its applications to homogeneous structures. Considering these disadvantages, the team decided to upgrade the sensor reported in [172] to a new and improved version based on a cantilever approach [179]. The fibre-optic nanomechanical probe (FONP) consisted of a microcantilever and a probe. The polymer pillar support was designed with a 20 μm width and a 40 μm height, and the cantilever was designed with a 60 μm length and a 20 μm width, while its thickness was varied using values of 6.3 μm (FONP-1), 2.8 μm (FONP-2), and 1.3 μm (FONP-3), allowing the researchers to identify the cantilever with the best performance. Two probes, one cylindrical and the other conical, were tested. After the TPP-DLW of the sensor structure, the upper surface of the microcantilever was gold-sputtered to enhance reflectivity. An FP cavity was formed between the end surface of the OF and the lower and upper surfaces of the microcantilever. When force was applied to the microcantilever, it deformed; consequently, the FP cavity was reduced, leading to a shorter OPL and changes in the optical spectrum. SEM images of the 3D-printed sensors, their respective spectral signatures, and the spectral characterisation results obtained for forces ranging from 0 to 35 μN are shown in figure 4.29.

In indentation tests on a glass substrate, the cantilever sensors with different thicknesses exhibited sensitivities of -0.5, -2.8, and -54.5 nm μN^{-1} for the FONP-1, FONP-2, and FONP-3 sensors, respectively. The sensitivity of the FONP-3 sensor was 30 times higher than that of the polymer clamped-beam probe proposed in their earlier study and discussed previously [172]. Furthermore, its detection limit was estimated to be 2.1 nN, comparable to the detection limit of AFM. The fabricated sensor was later tested by measuring the Young's modulus of soft materials, such as PDMS, onion cells, and human breast cancer cells, yielding successful results and validating the proposed sensor.

More recently, the fabrication of a spring-actuated FP sensor using the TPP method has been proposed to further reduce the detection limits of TPP-fabricated sensors to tens of piconewtons [180]. The cavity consisted of the DLW of a ring base, a triple helix, and a weighting cap at the tip of an SMF. The step-by-step fabrication and SEM images of the spring-actuated force sensors are shown in figure 4.30.

Due to the fragile nature of the structures, the team encountered difficulties in achieving a reliable spring-actuated FP sensor (see SEM image in figure 4.38(b)). However, by replacing isopropyl alcohol (IPA) with a lower-surface-tension solution of methyl nonafluorobutyl ether ($C_3H_5F_9O$) during the development stage, the authors successfully obtained a structure that was free of defects (see figure 4.30). The sensor's sensitivity was measured, reaching a value of (0.436 ± 0.007) nm nN^{-1}, enabling it to achieve a resolution of $\approx$40 pN. This resolution was one of the highest among most of the advanced force-sensing schemes based on OFs [173, 182] and MEMS sensors [182–184].

- **Vapour sensing**

The interaction of fibre-guided modes with the external environment is fundamental for monitoring the surrounding environment, whether in liquid or gaseous form. Fibre tapering (achieved by mechanically heat-deforming the fibre with CO_2 lasers or electric arc machines), chemical attack, and side-polishing procedures are ways to make the fundamental guided mode in an SMF interact with the external environment and thus sense the external medium. On the other hand, the use of technologies such as LPGs or multimode fibres in interferometric approaches, including

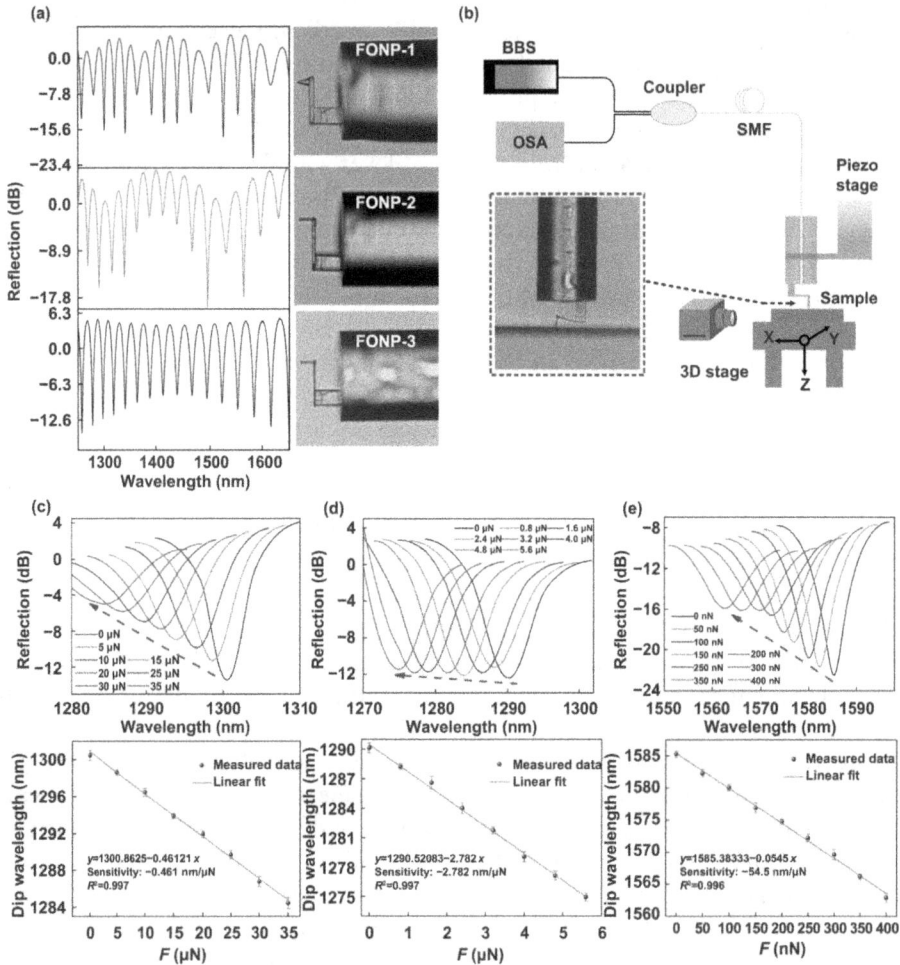

Figure 4.29. (a) Microscope images of the different FONPs and their respective spectra. (b) Representative schematic of the setup used for the force tests. Evolution of the sensor spectra and their respective dip wavelength evolution graphs for the sensors: (c) FONP-1, (d) FONP-2, and (e) FONP-3. Reproduced from [179]. © 2023 The Author(s). Published by IOP Publishing Ltd on behalf of the IMMT. CC BY 4.0.

multimode interferometers and MZI interferometers, is also a possible scenario. In contrast to a strong reliance on cumbersome fabrication methods and bulk fibre sensor technologies that always require transmission measurement schemes, coupled with the use of input and output fibre terminals, TPP introduces the possibility of developing tiny, microscopic sensing technologies at the tips of OFs, allowing the exploitation of lab-on-fibre concepts. For this, the use of technologies such as FP cavities, measured in reflection, can be explored at the tips of OFs through the development of specialised cavities that allow the light field to interact with the external environment, thereby measuring the surrounding refractive index. Even more interesting is the measurement of specific molecules of interest when these cavities are functionalised for a particular target molecule. The use of fibres with

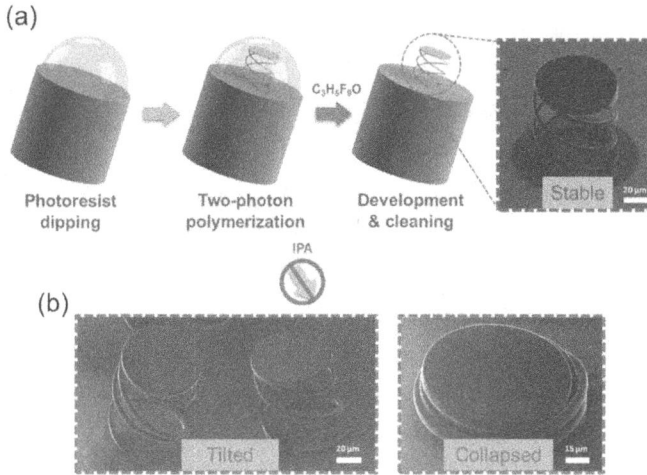

Figure 4.30. (a) Manufacturing process of the FP cavity, including photoresist fibre dipping, TPP printing, and development in methyl nonafluorobutyl. The result was a robust structure, as seen in the SEM image. (b) Damaged structure after photoresist development with isopropyl alcohol (IPA) [180] John Wiley & Sons. © 2023 Wiley-VCH GmbH.

multiple cores, namely multicore fibres (MCFs), is especially relevant in TPP sensor development. This occurs because the cores at the tip of an MCF can be accessed and used to directly print sensing structures that can be tailored for a specific sensor application or, in the fields of biology or chemistry, enable the detection of a specific target analyte and the full exploitation of lab-on-fibre technologies. Furthermore, the use of multiple cores enables the application of transmission detection schemes, where each fibre core can be treated as an input or an output as needed. However, to achieve such a configuration, the implementation of several optical components may be required, as is the case for prisms, tapers, waveguides, and other important optical structures. An example of this concept is illustrated in reference [185], where TPP-DLW was used to fabricate a high-quality-factor whispering-gallery micro-cavity (WGM) formed between two radial cores of a seven-core fibre for the detection of volatile organic compounds (VOCs). A representative schematic of this sensor, along with its SEM images, is displayed in figure 4.31.

The device comprised micropillars positioned at two outer cores, which served as light waveguides and supports for the structures anchored to them. At the tips of the pillars, two microprisms were used to reflect light in and out of two tapered waveguides anchored to those prisms. The tapered region of the waveguide was designed to couple light into two WGM rings of different radii and was positioned on top of two circular hollow truncated cones. This configuration allowed the researchers to obtain whispering-gallery resonant modes with high quality factors (i.e., 1.2×10^5 at ≈ 1540 nm), a characteristic that was only possible due to the low scattering losses of the WGM, which were ensured by its good surface quality, as displayed in figure 4.31.

WGM offers unique features, such as high temporal confinement of light and high sensitivity to the environment. A light beam can travel over 10^6 round trips inside a

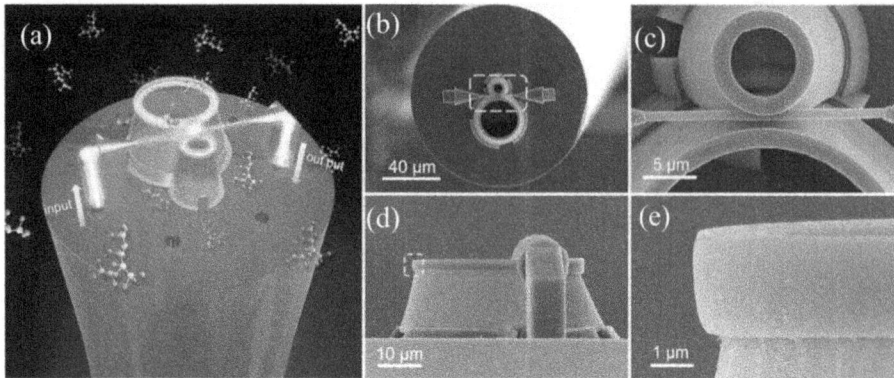

Figure 4.31. Schematic and SEM images of the WGM structures created to interconnect two diagonal cores for transmission interrogation. (a) Illustration of the light propagating through the waveguides, prisms, tapers, and ring waveguides interconnecting two MCF outer cores. The molecules spread around the MCF tip illustrate the target PGMEA molecules. SEM images of the TPP structure from top ((b) and (c)) and side ((d) and (e)) views. The images seen in (c) and (e) are close-ups of (b) and (d), respectively [185] John Wiley & Sons. © 2019 WILEY-VCH Verlag GmbH & Co. KGaA, Weinheim.

WGM, which significantly enhances light–matter interactions and enables a wide range of scientific discoveries and technological breakthroughs [186]. Because of these factors, the structure presented in [185] was used for the measurement of VOCs, such as propylene glycol monomethyl ether acetate (PGMEA), IPA, and alcohol (ALC). For this purpose, the fibre sensors were placed in a vapour environment produced for each of the mentioned VOC aqueous solutions. This interaction allowed the molecules to be absorbed by the WGM, which changed the ring radius due to swelling effects; in addition, the effective refractive index changed. Consequently, a change in spectral wavelength occurred. By monitoring the spectral resonance wavelengths, the authors were able to report sensitivities of 21.7, 3.4, and 3.9 pm ppm^{-1} for PGMEA, IPA, and ALC, respectively, in a low concentration range of less than 150 ppm, and with fast response times of 0.3 s.

- **Humidity**

In the literature, there have already been reports of the use of FP cavities fabricated through DLW for humidity sensor applications. This arose from the necessity in both industrial and scientific fields to measure humidity with high accuracy due to its influence on the physical and chemical properties of materials [187]. OF sensors are among the most extensively explored fibre sensors in this field [188]. Properties such as high sensitivity, low cost, remote measurement capability, and fast response are some of the advantages that have promoted the development of fibre-optic moisture sensors, including FBGs and LPGs [189, 190]. The compact size, simplicity, and high sensitivity of interferometers make them more appealing. Consequently, FP cavities have already been 3D printed on the tips of SMFs using a castle-like design [191]. The proposed FP cavity was directly manufactured on the tip of a cleaved OF. This was done by inserting a droplet of photoresist ($n = 1.52$) on its tip, followed by

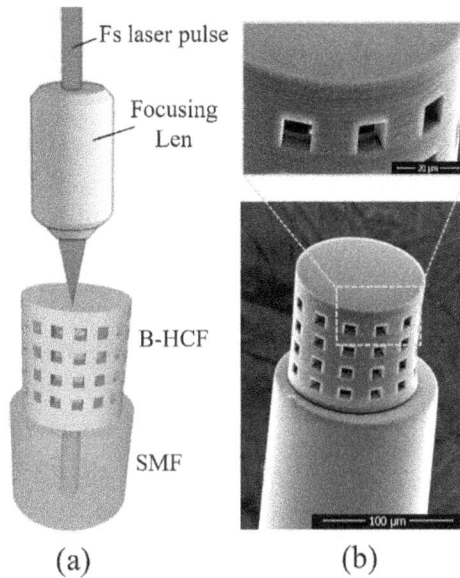

Figure 4.32. (a) Representative schematic (a) and SEM image (b) of the FP microcavity. Reprinted from [191], Copyright (2021), with permission from Elsevier.

fs DLW with a laser power of 20 mW and a writing speed of 1000 μm s^{-1}. The dimensions of the 3D-printed FP cavity were 100 μm for its length and 80 μm and 100 μm for its inner and outer diameters, respectively. A representative schematic of the 3D printing process and an SEM image of the printed microcavity are shown in figure 4.32.

To make the DLW-FP cavity sensitive to moisture, it was necessary to impregnate the cavity with a moisture-sensitive material, specifically polyvinyl alcohol (PVA). This was achieved by immersing the FP cavity in a PVA solution, allowing it to pass through the periodic, hollow square holes located on the side of the castle-like structure.

Tests for humidity sensing ranging from 46% to 75% revealed a spectral wavelength shift of the interference signal. The results showed good linearity and repeatability, presenting a sensitivity of ≈ -249 pm %RH^{-1}. Long-term moisture measurements, performed over 10 h, showed negligible changes of ≈ 40 pm. In addition, the temperature cross-sensitivity was evaluated by characterising the sensor at temperatures between 32.5 °C and 77.4 °C, resulting in a sensitivity of 28 pm °C^{-1}, leading to a low cross-sensitivity of 0.11 pm %RH^{-1}, showing that the sensor is a good candidate for high-performance humidity sensing.

- **Refractive index**

Fibre sensors at the tips of OFs are commonly fabricated using organic materials. Thus, they can absorb and swell in the presence of water, humidity [191], and VOCs [185]. While this feature has been used to measure these quantities, it can also lead to cross-sensitivity issues when other parameters need to be measured. Furthermore, the swelling and deswelling processes are time-dependent, relying on material properties and thickness, making this an undesirable feature in sensors. Additionally, the viscoelastic properties of polymers make them problematic in dynamic sensing schemes. Finally, the glass transition temperature of most polymers is below 300°C [192], which significantly limits the applications of polymer TPP fibre tip sensors. The 3D printing of inorganic materials has shown promising characteristics, namely good chemical resistance, thermal stability, optical transparency, and hardness. Therefore, the scientific community has already proposed the TPP fabrication of sensing structures such as FP cavities using inorganic glass [155]. For this, an HSQ solution was prepared and dropped on the tip of an OF. Later, the structure, consisting of a 3.5 μm thick plate suspended by four pillars at a distance of 5 μm from the fibre end, was engraved using DLW, which selectively polymerised the HSQ due to nonlinear absorption [193]. Then, the unexposed HSQ was removed through an in-house developing stage, and the structure was subsequently observed under SEM, as depicted in figures 4.33(a) and (b).

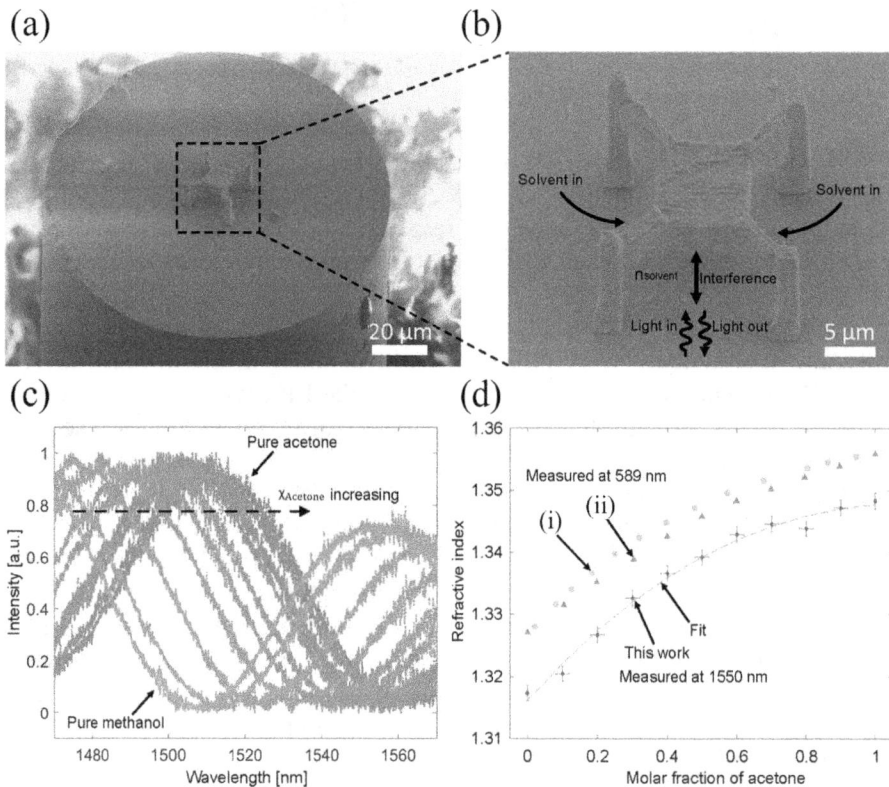

Figure 4.33. (a) Coloured SEM image of the FP cavity at the tip of the OF. (b) Close-up image of the FP cavity. (c) Normalised reflection spectra of the FP cavity for different molar fractions of acetone. (d) Measured refractive index as a function of the increase in the molar fraction of acetone. Some research results (i: [194] and ii: [195]) are also shown for comparison purposes. Adapted with permission from [155]. Copyright (2024) American Chemical Society.

The suspension of the 3.5 μm thick plate 5 μm above the fibre tip end face was designed to allow light exiting from the fibre core to interact with the surrounding environment. This allowed the authors to characterise the sensor for solutions with different refractive indices, specifically binary mixtures of acetone and methanol. Upon dipping the fibre tip sensor into the solutions, the authors were able to observe a red shift of the spectrum with an increasing molar fraction of acetone (0–1), as shown in figure 4.33(c). They were then able to extrapolate the refractive index of the sensor by measuring the wavelength shift. The results in figure 4.33(d) show that the refractive index followed a similar trend with increasing acetone molar fraction to those reported in the literature. The slightly lower values were attributed to dispersion, since the results presented in their study were obtained at 1550 nm, while the ones reported in the literature were measured at 589 nm.

References

[1] dos Santos J et al 2021 3D printing and nanotechnology: a multiscale alliance in personalized medicine Adv. Funct. Mater. 31 2009691
[2] Ee L Y and Yau Li S F 2021 Recent advances in 3D printing of nanocellulose: structure, preparation, and application prospects Nanoscale Adv. 3 1167–208
[3] Park Y G, Yun I, Chung W G, Park W, Lee D H and Park J U 2022 High-resolution 3D printing for electronics Adv. Sci. 9 2104623
[4] Park S, Shou W, Makatura L, Matusik W and (Kelvin) Fu K 2022 3D printing of polymer composites: materials, processes, and applications Matter 5 43–76
[5] Geisler E, Lecompère M and Soppera O 2022 3D printing of optical materials by processes based on photopolymerization: materials, technologies, and recent advances Photonics Res. 10 1344–60
[6] Zhiganshina E R et al 2022 Tetramethacrylic benzylidene cyclopentanone dye for one- and two-photon photopolymerization Eur. Polym. J. 162 110917
[7] Sun H B, Maeda M, Takada K, Chon J W M, Gu M and Kawata S 2003 Experimental investigation of single voxels for laser nanofabrication via two-photon photopolymerization Appl. Phys. Lett. 83 819–21
[8] Vaezi M, Seitz H and Yang S 2013 A review on 3D micro-additive manufacturing technologies Int. J. Adv. Manuf. Technol. 67 1721–54
[9] Berglund G, Wisniowiecki A, Gawedzinski J, Applegate B and Tkaczyk T S 2022 Additive manufacturing for the development of optical/photonic systems and components Optica 9 623–38
[10] Camposeo A, Persano L, Farsari M and Pisignano D 2019 Additive manufacturing: applications and directions in photonics and optoelectronics Adv. Opt. Mater. 7 1800419
[11] Mayer M G 1931 Über Elementarakte mit zwei Quantensprüngen Ann. Phys. 401 273–94
[12] Kaiser W and Garrett C G B 1961 Two-photon excitation in CaF_2: Eu^{2+} Phys. Rev. Lett. 7 229–31
[13] Zheng X, Cheng K, Zhou X, Lin J and Jing X 2017 A method for positioning the focal spot location of two photon polymerization AIP Adv. 7 095318
[14] Li L and Fourkas J T 2007 Multiphoton polymerization Mater. Today 10 30–7
[15] Skliutas E et al 2021 Polymerization mechanisms initiated by spatio-temporally confined light Nanophotonics 10 1211–42

[16] Farsari M, Vamvakaki M and Chichkov B N 2010 Multiphoton polymerization of hybrid materials *J. Opt.* **12** 124001

[17] Farsari M and Chichkov B N 2009 Materials processing: two-photon fabrication *Nat. Photonics* **3** 450–2

[18] Zhou W *et al* 2002 An efficient two-photon-generated photoacid applied to positive-tone 3D microfabrication *Science* **296** 1106–9

[19] Wu D *et al* 2009 100% Fill-factor aspheric microlens arrays (AMLA) with sub-20-nm precision *IEEE Photonics Technol. Lett.* **21** 1535–7

[20] Maruo S and Kawata S 1998 Two-photon-absorbed near-infrared photopolymerization for three-dimensional microfabrication *J. Microelectromech. Syst.* **7** 411–5

[21] Sun Z B, Dong X Z, Chen W Q, Nakanishi S, Duan X M and Kawata S 2008 Multicolor polymer nanocomposites: *in situ* synthesis and fabrication of 3D microstructures *Adv. Mater.* **20** 914–9

[22] Sun Z B, Dong X Z, Chen W Q, Shoji S, Duan X M and Kawata S 2008 Two- and three-dimensional micro/nanostructure patterning of CdS–polymer nanocomposites with a laser interference technique and *in situ* synthesis *Nanotechnology* **19** 035611

[23] Mendonca C R, Correa D S, Marlow F, Voss T, Tayalia P and Mazur E 2009 Three-dimensional fabrication of optically active microstructures containing an electroluminescent polymer *Appl. Phys. Lett.* **95** 113309

[24] Mizeikis V, Seet K K, Juodkazis S and Misawa H 2004 Three-dimensional woodpile photonic crystal templates for the infrared spectral range *Opt. Lett.* **29** 2061

[25] Aekbote B L, Fekete T, Jacak J, Vizsnyiczai G, Ormos P and Kelemen L 2016 Surface-modified complex SU-8 microstructures for indirect optical manipulation of single cells *Biomed. Opt. Express* **7** 45

[26] Dias G O, Lecarme O, Cordeiro J, Picard E and Peyrade D 2022 Microscale white light emitters fabricated by two-photon polymerization lithography on functional resist *Microelectron. Eng.* **257** p 111751

[27] Seiboth F *et al* 2022 Rapid aberration correction for diffractive X-ray optics by additive manufacturing *Opt. Express* **30** 31519–29

[28] Bajt S *et al* 2018 X-ray focusing with efficient high-NA multilayer Laue lenses *Light: Sci. Appl.* **7** 17162

[29] Seiboth F *et al* 2017 Perfect X-ray focusing via fitting corrective glasses to aberrated optics *Nat. Commun.* **8** 14623

[30] Yabashi M and Tanaka H 2017 The next ten years of X-ray science *Nat. Photonics* **11** 12–4

[31] Chao W *et al* 2012 Real space soft X-ray imaging at 10 nm spatial resolution *Opt. Express* **20** 9777–83

[32] Vila-Comamala J *et al* 2009 Advanced thin film technology for ultrahigh resolution X-ray microscopy *Ultramicroscopy* **109** 1360–4

[33] Evans-Lutterodt K *et al* 2003 Single-element elliptical hard X-ray micro-optics *Opt. Express* **11** 919–26

[34] Snigireva I *et al* 2001 Holographic X-ray optical elements: transition between refraction and diffraction *Nucl. Instrum. Methods Phys. Res. A.* **467** 982–5

[35] Sanli U T *et al* 2018 3D nanoprinted plastic kinoform X-ray optics *Adv. Mater.* **30** 1802503

[36] Jiang M *et al* 2021 3D high precision laser printing of a flat nanofocalizer for subwavelength light spot array *Opt. Lett.* **46** 356–9

[37] Lightman S, Bin-Nun M, Bar G, Hurvitz G and Gvishi R 2022 Structuring light using solgel hybrid 3D-printed optics prepared by two-photon polymerization *Appl. Opt.* **61** 1434–9

[38] Grier D G 2003 A revolution in optical manipulation *Nature* **424** 810–6

[39] Mair A, Vaziri A, Weihs G and Zeilinger A 2001 Entanglement of the orbital angular momentum states of photons *Nature* **412** 313–6

[40] Firth W J and Skryabin D V 1997 Optical solitons carrying orbital angular momentum *Phys. Rev. Lett.* **79** 2450–3

[41] Karimi E, Schulz S A, De Leon I, Qassim H, Upham J and Boyd R W 2014 Generating optical orbital angular momentum at visible wavelengths using a plasmonic metasurface *Light: Sci. Appl.* **3** e167

[42] Simpson N B, Dholakia K, Allen L, Padgett M J and Allen J F 1997 Mechanical equivalence of spin and orbital angular momentum of light: an optical spanner *Opt. Lett.* **22** 52–4

[43] Yan Y *et al* 2014 High-capacity millimetre-wave communications with orbital angular momentum multiplexing *Nat. Commun.* **5** 4876

[44] Berkhout G C G, Lavery M P J, Courtial J, Beijersbergen M W and Padgett M J 2010 Efficient sorting of orbital angular momentum states of light *Phys. Rev. Lett.* **105** 153601

[45] Lavery M P J *et al* 2012 Refractive elements for the measurement of the orbital angular momentum of a single photon *Opt. Express* **20** 2110–5

[46] Ruffato G, Massari M, Parisi G and Romanato F 2017 Test of mode-division multiplexing and demultiplexing in free-space with diffractive transformation optics *Opt. Express* **25** 7859–68

[47] Lightman S, Hurvitz G, Gvishi R and Arie A 2017 Miniature wide-spectrum mode sorter for vortex beams produced by 3D laser printing *Optica* **4** 605–10

[48] Ostrovsky A S, Rickenstorff-Parrao C and Arrizón V 2013 Generation of the 'perfect' optical vortex using a liquid-crystal spatial light modulator *Opt. Lett.* **38** 534–6

[49] Kotlyar V V, Almazov A A, Khonina S N, Soifer V A, Elfstrom H and Turunen J 2005 Generation of phase singularity through diffracting a plane or Gaussian beam by a spiral phase plate *J. Opt. Soc. Am.* A **22** 849–60

[50] Zhou J and Lin P T 2022 Generation of mid-infrared vortex beams by 3-D printed polymer phase plates *Opt. Laser Technol.* **156** 108509

[51] Yu J *et al* 2020 3D nanoprinted kinoform spiral zone plates on fiber facets for high-efficiency focused vortex beam generation *Opt. Express* **28** 38127–39

[52] Wang J, Cai C, Wang K and Wang J 2022 Generation of bessel beams via femtosecond direct laser writing 3D phase plates *Opt. Lett.* **47** 5766–9

[53] Lightman S, Porat O, Hurvitz G and Gvishi R 2022 Vortex-bessel beam generation by 3D direct printing of an integrated multi-optical element on a fiber tip *Opt. Lett.* **47** 5248–51

[54] Goi E, Schoenhardt S and Gu M 2022 Direct retrieval of Zernike-based pupil functions using integrated diffractive deep neural networks *Nat. Commun.* **13** 7531

[55] Nussbaum P, Völkel R, Herzig H P, Eisner M and Haselbeck S 1997 Design, fabrication and testing of microlens arrays for sensors and microsystems *Pure Appl. Opt.* **6** 617–36

[56] Yao J *et al* 2001 Refractive micro lens array made of dichromate gelatin with gray-tone photolithography *Microelectron. Eng.* **57** 729–35

[57] He M, Yuan X C, Ngo N Q, Bu J and Tao S H 2004 Single-step fabrication of a microlens array in sol-gel material by direct laser writing and its application in optical coupling *J. Opt. A Pure Appl. Opt.* **6** 94–7

[58] Sugiyama S, Khumpuang S and Kawaguchi G 2004 Plain-pattern to cross-section transfer (PCT) technique for deep X-ray lithography and applications *J. Micromech. Microeng.* **14** 1399–404

[59] Serbin J and Chichkov B N 2004 High-resolution direct-write femtosecond laser technologies *Solid State Laser Technologies and Femtosecond Phenomena* (Bellingham, WA: SPIE) pp. 245–51

[60] Guo R, Xiao S, Zhai X, Li J, Xia A and Huang W 2006 Micro lens fabrication by means of femtosecond two photon photopolymerization *Opt. Express* **14** 810–6

[61] Xu J J *et al* 2015 High curvature concave-convex microlens *IEEE Photonics Technol. Lett.* **27** 2465–8

[62] Popovic Z D, Sprague R A and Connell G A N 1988 Technique for monolithic fabrication of microlens arrays *Appl. Opt.* **27** 1281–4

[63] Zeng X *et al* 2002 A simple method for microlens fabrication by the modified LIGA process *J. Micromech. Microeng.* **12** 334–40

[64] Lee B K, Kim D S and Kwon T H 2004 Replication of microlens arrays by injection molding *Microsyst.* **10** 531–5

[65] Kim J Y *et al* 2011 Hybrid polymer microlens arrays with high numerical apertures fabricated using simple ink-jet printing technique *Opt. Mater. Exoress.* **1** 259–69

[66] Völkel R, Eisner M and Weible K J 2003 Miniaturized imaging systems *Microelectron. Eng.* **67–8** 461–72

[67] Kuang D, Zhang X, Gui M and Fang Z 2009 Hexagonal microlens array fabricated by direct laser writing and inductively coupled plasma etching on organic light emitting devices to enhance the outcoupling efficiency *Appl. Opt.* **48** 974–8

[68] Yang R, Wang W and Soper S A 2005 Out-of-plane microlens array fabricated using ultraviolet lithography *Appl. Phys. Lett.* **86** 1–3

[69] Hoy C L *et al* 2008 Miniaturized probe for femtosecond laser microsurgery and two-photon imaging *Opt. Express* **16** 9996–10005

[70] Gissibl T, Thiele S, Herkommer A and Giessen H 2016 Two-photon direct laser writing of ultracompact multi-lens objectives *Nat. Photonics* **10** 554–60

[71] Ye M and Sato S 2002 Optical properties of liquid crystal lens of any size *Jpn. J. Appl. Phys.* **41** 571–3

[72] Ren H, Fan Y H, Gauza S and Wu S T 2004 Tunable microlens arrays using polymer network liquid crystal *Opt. Commun.* **230** 267–71

[73] Asatryan K, Presnyakov V, Tork A, Zohrabyan A, Bagramyan A and Galstian T 2010 Optical lens with electrically variable focus using an optically hidden dielectric structure *Opt. Express* **18** 13981–92

[74] Commander L G, Day S E and Selviah D R 2000 Variable focal length microlenses *Opt. Commun.* **177** 157–70

[75] He Z, Lee Y-H, Chanda D and Wu S-T 2018 Adaptive liquid crystal microlens array enabled by two-photon polymerization *Opt. Express* **26** 21184–93

[76] Tanida J *et al* 2001 Thin observation module by bound optics TOMBO: concept and experimental verification *Appl. Opt.* **40** 1806–13

[77] Hu Z Y *et al* 2022 Miniature optoelectronic compound eye camera *Nat. Commun.* **13** 5634

[78] Dai B *et al* 2021 Biomimetic apposition compound eye fabricated using microfluidic-assisted 3D printing *Nat. Commun.* **12** 6458

[79] Smolyaninov I I, Hung Y J and Davis C C 2007 Magnifying superlens in the visible frequency range *Science* **315** 1699–701

[80] Hecht B *et al* 2000 Scanning near-field optical microscopy with aperture probes: fundamentals and applications *J. Chem. Phys.* **112** 7761–74

[81] Hell S W 1979 Far-field optical nanoscopy *Science* **316** 1153–8 2007

[82] Wang Z *et al* 2011 Optical virtual imaging at 50 nm lateral resolution with a white-light nanoscope *Nat. Commun.* **2** 218

[83] Zhu H, Chen M, Zhou S and Wu L 2017 Synthesis of high refractive index and shape controllable colloidal polymer microspheres for super-resolution imaging *Macromolecules* **50** 660–5

[84] Du B, Zhang H, Xia J, Wu J, Ding H and Tong G 2020 Super-resolution imaging with direct laser writing-printed microstructures *J. Phys. Chem.* A **124** 7211–6

[85] Žukauskas A, Matulaitiene I, Paipulas D, Niaura G, Malinauskas M and Gadonas R 2015 Tuning the refractive index in 3D direct laser writing lithography: towards GRIN microoptics *Laser. Photon. Rev.* **9** 706–12

[86] Ocier C R *et al* 2020 Direct laser writing of volumetric gradient index lenses and waveguides *Light: Sci. Appl.* **9** 196

[87] Li J *et al* 2020 Ultrathin monolithic 3D printed optical coherence tomography endoscopy for preclinical and clinical use *Light: Sci. Appl.* **9** 124

[88] Dietrich P I *et al* 2018 In situ 3D nanoprinting of free-form coupling elements for hybrid photonic integration *Nat. Photonics* **12** 241–7

[89] Huang P-H *et al* 2023 Three-dimensional printing of silica glass with sub-micrometer resolution *Nat. Commun.* **14** 3305

[90] Hong Z, Ye P, Loy D A and Liang R 2021 Three-dimensional printing of glass micro-optics *Optica* **8** 904–10

[91] Hong Z, Ye P, Loy D A and Liang R 2022 High-precision printing of complex glass imaging optics with precondensed liquid silica resin *Adv. Sci.* **9** 2105595

[92] Khorasaninejad M, Chen W T, Devlin R C, Oh J, Zhu A Y and Capasso F 2016 Metalenses at visible wavelengths: diffraction-limited focusing and subwavelength resolution imaging *Science* **352** 1190–4

[93] Devlin R C, Ambrosio A, Rubin N A, Mueller J P B and Capasso F 2017 Arbitrary spin-to-orbital angular momentum conversion of light *Science* **358** 896–901

[94] Ren H *et al* 2019 Metasurface orbital angular momentum holography *Nat. Commun.* **10** 2986

[95] Roques-Carmes C *et al* 2022 Toward 3D-printed inverse-designed metaoptics *ACS Photonics* **9** 43–51

[96] Liberale C *et al* 2010 Micro-optics fabrication on top of optical fibers using two-photon lithography *IEEE Photonics Technol. Lett.* **22** 474–6

[97] Yuan G H, Rogers E T and Zheludev N I 2017 Achromatic super-oscillatory lenses with sub-wavelength focusing *Light: Sci. Appl.* **6** e17036

[98] Kuchmizhak A, Gurbatov S, Nepomniaschii A, Vitrik O and Kulchind Y 2014 High-quality fiber microaxicons fabricated by a modified chemical etching method for laser focusing and generation of Bessel-like beams *Appl. Opt.* **53** 937–43

[99] Consales M, Ricciardi A, Crescitelli A, Esposito E, Cutolo A and Cusano A 2012 Lab-on-fiber technology: toward multifunctional optical nanoprobes *ACS Nano* **6** 3163–70

[100] Kostovski G, Chinnasamy U, Jayawardhana S, Stoddart P R and Mitchell A 2011 Sub-15nm optical fiber nanoimprint lithography: a parallel, self-aligned and portable approach *Adv. Mater.* **23** 531–5

[101] Ren H *et al* 2022 An achromatic metafiber for focusing and imaging across the entire telecommunication range *Nat. Commun.* **13** 4183

[102] Lin C *et al* 2018 Optofluidic gutter oil discrimination based on a hybrid-waveguide coupler in fibre *Lab. Chip.* **18** 595–600

[103] Zhao J *et al* 2016 Surface plasmon resonance refractive sensor based on silver-coated side-polished fiber *Sensors Actuators* B **230** 206–11

[104] Li Z *et al* 2017 Label-free detection of bovine serum albumin based on an in-fiber Mach–Zehnder interferometric biosensor *Opt. Express* **25** 17105–13

[105] Presby H M, Benner A F and Edwards C A 1990 Laser micromachining of efficient fiber microlenses *Appl. Opt.* **29** 2692–5

[106] Schiappelli F *et al* 2004 Efficient fiber-to-waveguide coupling by a lens on the end of the optical fiber fabricated by focused ion beam milling *Microelectron. Eng.* **73** 397–404

[107] Cojoc G *et al* 2010 Optical micro-structures fabricated on top of optical fibers by means of two-photon photopolymerization *Microelectron. Eng.* **87** 876–9

[108] Gissibl T, Thiele S, Herkommer A and Giessen H 2016 Sub-micrometre accurate free-form optics by three-dimensional printing on single-mode fibres *Nat. Commun.* **7** 11763

[109] Gissibl T, Schmid M and Giessen H 2016 Spatial beam intensity shaping using phase masks on single-mode optical fibers fabricated by femtosecond direct laser writing *Optica* **3** 448–51

[110] Xie Z *et al* 2015 Demonstration of a 3D radar-like SERS sensor micro- and nanofabricated on an optical fiber *Adv. Opt. Mater.* **3** 1232–9

[111] Park H, Kim S, Park J, Joo J and Kim G 2013 A fiber-to-chip coupler based on Si/SiON cascaded tapers for Si photonic chips *Opt. Express* **21** 29313–9

[112] Ren G, Chen S, Cheng Y and Zhai Y 2011 Study on inverse taper based mode transformer for low loss coupling between silicon wire waveguide and lensed fiber *Opt. Commun.* **284** 4782–8

[113] Tiecke T G *et al* 2015 Efficient fiber-optical interface for nanophotonic devices *Optica* **2** 70–5

[114] Chang L *et al* 2015 Waveguide-coupled micro-ball lens array suitable for mass fabrication *Opt. Express.* **23** 22414–23

[115] Michaels A and Yablonovitch E 2018 Inverse design of near unity efficiency perfectly vertical grating couplers *Opt. Express* **26** 4766–79

[116] Gehring H *et al* 2019 Low-loss fiber-to-chip couplers with ultrawide optical bandwidth *APL Photonics* **4** 010801

[117] Vanmol K, Tuccio S, Panapakkam V, Thienpont H, Watté J and Van Erps J 2019 Two-photon direct laser writing of beam expansion tapers on single-mode optical fibers *Opt. Laser Technol.* **112** 292–8

[118] Vanmol K *et al* 2020 Mode-field matching down-tapers on single-mode optical fibers for edge coupling towards generic photonic integrated circuit platforms *J. Lightwave Technol.* **38** 4834–42

[119] Vanmol K *et al* 2020 3D direct laser writing of microstructured optical fiber tapers on single-mode fibers for mode-field conversion *Opt. Express* **28** 36147–58

[120] Russell P 2003 Photonic crystal fibers *Science* **299** 358–62

[121] Bjarklev A, Broeng J and Bjarklev A S 2003 Fabrication of photonic crystal fibers in *Photonic Crystal Fibres* (Berlin: Springer) pp. 115–30

[122] Peng G-D *et al* 2019 3D silica lithography for future optical fiber fabrication in *Handbook of Optical Fibers* (Berlin: Springer) pp. 1–17

[123] Cook K *et al* 2015 Air-structured optical fiber drawn from a 3D-printed preform *Opt. Lett.* **40** 3966

[124] Talataisong W *et al* 2018 Mid-IR hollow-core microstructured fiber drawn from a 3D printed PETG preform *Sci Rep.* **8** 8113

[125] Bertoncini A and Liberale C 2020 3D printed waveguides based on photonic crystal fiber designs for complex fiber-end photonic devices *Optica* **7** 1487–94

[126] Malinauskas M *et al* 2012 3D microoptical elements formed in a photostructurable germanium silicate by direct laser writing *Opt. Lasers Eng.* **50** 1785–8

[127] Bianchi S, Rajamanickam V P, Ferrara L, Di Fabrizio E, Liberale C and Di Leonardo R 2013 Focusing and imaging with increased numerical apertures through multimode fibers with micro-fabricated optics *Opt. Lett.* **38** 4935–8

[128] Trappen M *et al* 2020 3D-printed optical probes for wafer-level testing of photonic integrated circuits *Opt. Express* **28** 37996–8007

[129] Kretschmann E and Raether H 1968 Radiative decay of non radiative surface plasmons excited by light *Z. Naturforsch.* A **23** 2135–6

[130] Otto A 1968 Excitation of nonradiative surface plasma waves in silver by the method of frustrated total reflection *Z. Phys.* **216** 398–410

[131] Safronov K R *et al* 2022 Miniature otto prism coupler for integrated photonics *Laser Photonics Rev.* **16** 2100542

[132] Kovalevich T *et al* 2017 Polarization controlled directional propagation of Bloch surface wave *Opt. Express* **25** 5710–5

[133] Wang X *et al* 2011 Enhanced cell sorting and manipulation with combined optical tweezer and microfluidic chip technologies *Lab Chip* **11** 3656–62

[134] Eriksson E *et al* 2007 A microfluidic system in combination with optical tweezers for analyzing rapid and reversible cytological alterations in single cells upon environmental changes *Lab Chip* **7** 71–6

[135] Lincoln B *et al* 2007 Reconfigurable microfluidic integration of a dual-beam laser trap with biomedical applications *Biomed. Microdevices* **9** 703–10

[136] Cran-McGreehin S, Krauss T F and Dholakia K 2006 Integrated monolithic optical manipulation *Lab Chip* **6** 1122–4

[137] Kawata S and Tani T 1996 Optically driven Mie particles in an evanescent field along a channeled waveguide *Opt. Lett.* **21** 1768–70

[138] Juan M L, Righini M and Quidant R 2011 Plasmon nano-optical tweezers *Nat. Photonics* **5** 349–56

[139] Sun Y Y, Yuan X C, Ong L S, Bu J, Zhu S W and Liu R 2007 Large-scale optical traps on a chip for optical sorting *Appl. Phys. Lett.* **90** 031107

[140] Yu S *et al* 2021 On-chip optical tweezers based on freeform optics *Optica* **8** 409–14

[141] Yang A H J, Lerdsuchatawanich T and Erickson D 2009 Forces and transport velocities for a particle in a slot waveguide *Nano Lett.* **9** 1182–8

[142] Kim J and Shin J H 2016 Stable, free-space optical trapping and manipulation of sub-micron particles in an integrated microfluidic chip *Sci Rep.* **6** 33842

[143] Bertoncini A, Cojoc G, Guck J and Liberale C 2020 High-throughput fabrication of right-angle prism mirrors with selective metalization by two-step 3D printing and computer vision alignment *Proceedings SPIE* (Bellingham, WA: SPIE-International Society of Optical Engineering) p 1129211

[144] Liberale C *et al* 2013 Integrated microfluidic device for single-cell trapping and spectroscopy *Sci Rep.* **3** 1258

[145] Atwater J H *et al* 2011 Microphotonic parabolic light directors fabricated by two-photon lithography *Appl. Phys. Lett.* **99** 151113

[146] Williams J, Smith J, Suelzer J S, Usechak N G and Chandrahalim H 2020 Optical fiber-tip heat sensor featuring a multipositional fabry-pérot cavity resonator *IEEE Sensors* **2020** 1–4

[147] Smith J W, Suelzer J S, Usechak N G, Tondiglia V P and Chandrahalim H 2019 3-D thermal radiation sensors on optical fiber tips fabricated using ultrashort laser pulses *Transducers 2019—Eurosensors XXXIII* 649–52

[148] Williams J C, Chandrahalim H, Suelzer J S and Usechak N G 2022 Multiphoton nanosculpting of optical resonant and nonresonant microsensors on fiber tips *ACS Appl. Mater. Interfaces* **14** 19988–99

[149] Kostovski G, Stoddart P R and Mitchell A 2014 The optical fiber tip: an inherently light-coupled microscopic platform for micro- and nanotechnologies *Adv. Mater.* **26** 3798–820

[150] Vanek M, Vanis J, Baravets Y, Todorov F, Ctyroky J and Honzatko P 2016 High-power fiber laser with a polarizing diffraction grating milled on the facet of an optical fiber *Opt. Express* **24** 30225

[151] Hahn V, Kalt S, Sridharan G M, Wegener M and Bhattacharya S 2018 Polarizing beam splitter integrated onto an optical fiber facet *Opt. Express* **26** 33148

[152] Wei H *et al* 2019 Two-photon direct laser writing of inverse-designed free-form near-infrared polarization beamsplitter *Adv. Opt. Mater.* **7** 1900513

[153] Jiang H *et al* 2014 Polarization splitter based on dual-core photonic crystal fiber *Opt. Express* **22** 30461–6

[154] Kotz F *et al* 2017 Three-dimensional printing of transparent fused silica glass *Nature* **544** 337–9

[155] Lai L L, Huang P H, Stemme G, Niklaus F and Gylfason K B 2023 3D printing of glass micro-optics with subwavelength features on optical fiber tips *ACS Nano* **18** 10788–97

[156] Bertoncini A and Liberale C 2018 Polarization micro-optics: circular polarization from a fresnel rhomb 3D printed on an optical fiber *IEEE Photonics Technol. Lett.* **30** 1882–5

[157] Lin T *et al* 1966 To cite this article: R J King *J. Sci. Instrum.* **43** 617–20

[158] Dzibrou D O, Van Der Tol J J G M and Smit M K 2013 Improved fabrication process of low-loss and efficient polarization converters in InP-based photonic integrated circuits *Opt. Lett.* **38** 1061–3

[159] Zhang H *et al* 2012 Efficient and broadband polarization rotator using horizontal slot waveguide for silicon photonics *Appl. Phys. Lett.* **101** 021105

[160] Caspers J N, Alam M Z and Mojahedi M 2012 Compact hybrid plasmonic polarization rotator *Opt. Lett.* **37** 4615–7

[161] Hou Z S *et al* 2019 On-chip polarization rotators *Adv. Opt. Mater.* **7** 1900129

[162] Nesic A *et al* 2023 Ultra-broadband polarization beam splitter and rotator based on 3D-printed waveguides *Light Adv. Manuf.* **4** 251–62

[163] Warner M and Terentjev E M 2003 *Liquid Crystal Elastomers* **120** (Oxford: Oxford University Press)

[164] Xiang X, Kim J, Komanduri R and Escuti M J 2017 Nanoscale liquid crystal polymer Bragg polarization gratings *Opt. Express* **25** 19298–308

[165] Nocentini S, Martella D, Parmeggiani C, Zanotto S and Wiersma D S 2018 Structured optical materials controlled by light *Adv. Opt. Mater.* **6** 1800167

[166] Akamatsu N, Hisano K, Tatsumi R, Aizawa M, Barrett C J and Shishido A 2017 Thermo-, photo-, and mechano-responsive liquid crystal networks enable tunable photonic crystals *Soft Matter* **13** 7486–91

[167] Zanotto S, Sgrignuoli F, Nocentini S, Martella D, Parmeggiani C and Wiersma D S 2019 Multichannel remote polarization control enabled by nanostructured liquid crystalline networks *Appl. Phys. Lett.* **114** 201103

[168] Barkley D, Song B, Mukund V, Lemoult G, Avila M and Hof B 2015 The rise of fully turbulent flow *Nature* **526** 550–3

[169] Hof B, De Lozar A, Avila M, Tu X, Tobias † and Schneider M 2010 Eliminating turbulence in spatially intermittent flows *Science* **327** 1491–4

[170] Martínez-Martín D *et al* 2017 Inertial picobalance reveals fast mass fluctuations in mammalian cells *Nature* **550** 500–5

[171] Cermak N *et al* 2016 High-throughput measurement of single-cell growth rates using serial microfluidic mass sensor arrays *Nat. Biotechnol.* **34** 1052–9

[172] Zou M *et al* 2021 Fiber-tip polymer clamped-beam probe for high-sensitivity nanoforce measurements *Light: Sci. Appl.* **10** 171

[173] Wu Y *et al* 2018 Highly sensitive force sensor based on balloon-like interferometer *Opt. Laser Technol.* **103** 17–21

[174] Liu Q, Xing L, Wu Z, Cai L, Zhang Z and Zhao J 2020 High-sensitivity photonic crystal fiber force sensor based on Sagnac interferometer for weighing *Opt. Laser Technol.* **123** 105939

[175] Chung K M, Liu Z, Lu C and Tam H Y 2012 Highly sensitive compact force sensor based on microfiber bragg grating *IEEE Photonics Technol. Lett.* **24** 700–2

[176] Liu Y, Qu S, Qu W and Que R 2014 A Fabry-Perot cuboid cavity across the fibre for high-sensitivity strain force sensing *J. Opt.* **16** 105401

[177] Gong Y *et al* 2014 Highly sensitive force sensor based on optical microfiber asymmetrical Fabry-Perot interferometer *Opt. Express* **22** 3578–84

[178] Liu Y, Lang C, Wei X and Qu S 2017 Strain force sensor with ultra-high sensitivity based on fiber inline Fabry–Perot micro-cavity plugged by cantilever taper *Opt. Express* **25** 7797–806

[179] Zou M *et al* 2023 3D printed fiber-optic nanomechanical bioprobe *Int. J. Extreme Manuf.* **5** 015005

[180] Shang X *et al* 2024 Fiber-integrated force sensor using 3D printed spring-composed Fabry-Perot cavities with a high precision down to tens of piconewton *Adv. Mater.* **36** 2305121

[181] Power M, Thompson A J, Anastasova S and Yang G Z 2018 A monolithic force-sensitive 3D microgripper fabricated on the tip of an optical fiber using 2-photon polymerization *Small* **14** 1703964

[182] Wei Y and Xu Q 2017 Design of a PVDF-MFC force sensor for robot-assisted single cell microinjection *IEEE Sens. J.* **17** 3975–82

[183] Zhang W *et al* 2016 Design and characterization of a novel T-shaped multi-axis piezoresistive force/moment sensor *IEEE Sens. J.* **16** 4198–210

[184] Nastro A, Ferrari M and Ferrari V 2020 Double-actuator position-feedback mechanism for adjustable sensitivity in electrostatic-capacitive MEMS force sensors *Sens. Actuators A Phys.* **312** 112127

[185] Zhang S *et al* 2019 High-Q polymer microcavities integrated on a multicore fiber facet for vapor sensing *Adv. Opt. Mater.* **7** 1900602

[186] Otuka A J G, Tomazio N B, Paula K T and Mendonça C R 2021 Two-photon polymerization: functionalized microstructures, micro-resonators, and bio-scaffolds *Polymers (Basel)* **13** 1994

[187] Choi S J *et al* 2018 Nitrogen-doped single graphene fiber with platinum water dissociation catalyst for wearable humidity sensor *Small* **14** 1703934

[188] Chen M-Q, Zhao Y, Xia F, Peng Y and Tong R-J 2018 High sensitivity temperature sensor based on fiber air-microbubble Fabry-Perot interferometer with PDMS-filled hollow-core fiber *Sens. Actuators A Phys.* **275** 60–6

[189] Yan G, Liang Y, Lee E-H and He S 2015 Novel Knob-integrated fiber Bragg grating sensor with polyvinyl alcohol coating for simultaneous relative humidity and temperature measurement *Opt. Express* **23** 15624–34

[190] Liu Y, Wang L, Zhang M, Tu D, Mao X and Liao Y 2007 Long-period grating relative humidity sensor with hydrogel coating *IEEE Photonics Technol. Lett.* **19** 880–2

[191] Chen M-Q, Zhao Y, Wei H-M, Zhu C-L and Krishnaswamy S 2021 3D printed castle style Fabry–Perot microcavity on optical fiber tip as a highly sensitive humidity sensor *Sensors Actuators* B **328** 128981

[192] Mark J E 2006 *Physical Properties of Polymers Handbook* 2nd edn (Berlin: Springer)

[193] Huang P H *et al* 2023 Three-dimensional printing of silica glass with sub-micrometer resolution *Nat. Commun.* **14** 3305

[194] Iglesias M, Orge B, Domínguez M and Tojo J 1998 Mixing properties of the binary mixtures of acetone, methanol, ethanol, and 2-butanone at 298.15 K *Phys. Chem. Liq.* **37** 9–29

[195] Mohammadi M D and Hamzehloo M 2019 Densities, viscosities, and refractive indices of binary and ternary mixtures of methanol, acetone, and chloroform at temperatures from (298.15–318.15) K and ambient pressure *Fluid Phase Equilib.* **483** 14–30

www.ingramcontent.com/pod-product-compliance
Lightning Source LLC
Chambersburg PA
CBHW080546220326
41599CB00032B/6376